AMERICA'S HORROR STORIES

America's Horror Stories: U.S. History through Dark Tourism conducts a ghost tour(ist) methodology to explore how slavery and racism are represented in dark tourism via ghost tours.

The authors travel to key sites of racist U.S. history, including Salem, Massachusetts, where a witch panic was sparked by accusations of witchcraft by Tituba, an enslaved woman practicing Voodoo; New Orleans, Louisiana, which hosts the largest slave trade market; the Myrtles Plantation in Francisville, Louisiana; and to Gettysburg, Pennsylvania, where the bloodiest battle of the Civil War took place, marking a pivotal moment to end slavery in the nation—but where Confederate ghosts are said to continue roaming the town and battlefield. Acting as research ghost hunters/tourists, the authors go on walking and bus tours, visit historical monuments, stay at haunted hotels, ponder objects in haunted museums, and do some ghost hunting of their own. They find that the ghosts conjured by tour guides—ghosts of Confederate soldiers, American citizens, and enslaved people—tend to whitewash, sensationalize, and commercialize the horrors of U.S. history, including slavery, racism, and colonialism. They do not discount dark tourism entirely, but recommend a ghost tour(ist) pedagogy that critically considers social issues—and structural forms of inequality—that haunt us today.

America's Horror Stories will be of great interest to students and scholars researching and taking part in critical criminology and cultural criminology courses, specifically on crime, media, and culture.

Kevin Revier is an assistant professor in the Sociology/Anthropology Department at SUNY Cortland, U.S.A.

Favian Alejandro Martín is an associate professor in the Department of Sociology, Anthropology, and Criminal Justice at Arcadia University, U.S.A.

ROUTLEDGE STUDIES IN CRIME, CULTURE AND MEDIA

Routledge Studies in Crime, Culture and Media offers the very best in research that seeks to understand crime through the context of culture, cultural processes and media.

The series welcomes monographs and edited volumes from across the globe, and across a variety of disciplines. Books will offer fresh insights on a range of topics, including news reporting of crime; moral panics and trial by media; media and the police; crime in film; crime in fiction; crime in TV; crime and music; 'reality' crime shows; the impact of new media including mobile, Internet and digital technologies, and social networking sites; the ways media portrayals of crime influence government policy and lawmaking; the theoretical, conceptual and methodological underpinnings of cultural criminology.

Books in the series will be essential reading for those researching and studying criminology, media studies, cultural studies and sociology.

Social Media and Law Enforcement Practice in Poland
Insights into Practice Outside Anglophone Countries
Edited by Paweł Waszkiewicz

True Crime and Women
Writers, Readers, and Representations
Edited by Lili Pâquet and Rose Williamson

Social Media and Criminal Justice
Xiaochen Hu and Nicholas P. Lovrich

A Popular Criminology of Youth Justice
Youth on Film
Jessica Urwin

America's Horror Stories
U.S. History through Dark Tourism
Kevin Revier and Favian Alejandro Martín

AMERICA'S HORROR STORIES

U.S. History through Dark Tourism

Kevin Revier and Favian Alejandro Martín

Routledge
Taylor & Francis Group

LONDON AND NEW YORK

Designed cover image: © JGregorySF / Getty Images

First published 2025
by Routledge
4 Park Square, Milton Park, Abingdon, Oxon OX14 4RN

and by Routledge
605 Third Avenue, New York, NY 10158

Routledge is an imprint of the Taylor & Francis Group, an informa business

British Library Cataloguing-in-Publication Data
A catalogue record for this book is available from the British Library

ISBN: 978-1-032-50295-3 (hbk)
ISBN: 978-1-032-50292-2 (pbk)
ISBN: 978-1-003-39780-9 (ebk)

DOI: 10.4324/9781003397809

Typeset in Sabon
by Taylor & Francis Books

I would like to dedicate this book to my family and friends, Carol, Doug, Amy, Benny, Indigo, and Olivia—and to Nate, who I would never ghost.

I would like to dedicate this book to my parents, Jen and Carlos Martín, for their unwavering support in everything I do. Without their support and encouragement, this book would not have happened.

CONTENTS

FIGURES

ACKNOWLEDGMENT

First and foremost, we would like to thank the Center for Antiracist Scholarship, Advocacy, and Action (CASAA) at Arcadia University for providing us with the resources needed to complete this research through the W.K. Kellogg Foundation. We are very grateful to Drs. Doreen Loury (CASAA Founding Director) and Christopher Varlack (CASAA Executive Director) for supporting us in this research endeavor. We also want to thank the Sociology, Anthropology, and Criminal Justice Department as well as our research assistants, Justin Soto, Kyle Stump, Emily Doorbejai, and Connor Wright. Without their assistance, we could not have completed this book.

INTRODUCTION

Ghosting Dark Tourism

Introduction

"Not your average guest selfie," read a December 4, 2017, Facebook post uploaded by the Myrtles Plantation, a slave plantation turned bed and breakfast in St. Francisville, Louisiana. The post went viral. It garnered over 3,000 "likes" and 2,500 "shares," and it received coverage by *Good Morning America* (2017). The photo features six white women smiling. But something is unusual. In the window behind them appears the ghostly figure of a Black woman. Many wonder, a reporter explains, if she "is Chloe, a slave girl who some say has been in several photographs taken at the plantation" (WDSU, 2021).

As is evident, ghosts are big business. Three out of four Americans believe in the paranormal (Luhrmann, 2013) and slightly over 40 percent agree places can be haunted by spirits (Bader, et al., 2017, 103). The burgeoning ghost tourism industry caters to such beliefs, as guides lead visitors through cemeteries, mansions, and even former slave plantations (Becker, 2013). An abundance of television series sustains a collective fascination with ghosts, including SYFY's *Ghost Hunters* (2003–Present), Discovery's *Ghost Lab* (2009–2011), A&E's *Extreme Paranormal* (2009) and *Paranormal State* (2007–2011), and Travel Channel's *Most Haunted* (2014–2019) and *Ghost Adventures* (2008–Present; Bader et al., 2017, 3). Haunted attractions bring in approximately $300 million annually in the United States (Yuko, 2021). "4 Haunted Attractions That Will Push You to the Limits of Your Sanity!," the Pennsylvania Pennhurst Asylum in Spring City, Pennsylvania, advertises. Philadelphia's Eastern State Penitentiary, the first modern American prison which used isolation to "rehabilitate" those incarcerated, presents, "A Halloween Festival of Epic Proportions." Or, one may "Escape to Myrtles Plantation" to perhaps see a glimpse of Chloe, the enslaved ghost of Facebook fame.

DOI: 10.4324/9781003397809-1

Yet, ghost tours do more than offer a thrilling experience: they shape how we think about the past. As Nikole Hannah-Jones (2021, xxvi) reminds us,

> [W]hile history *is* what happened, it is also, just as important, how we *think* about what happened and what we unearth and choose to remember about what happened.

After all, Pennhurst Asylum, while a place of marketed thrill, engaged in eugenic experimentation and locked up those considered "imbeciles" or "insane". Eastern State Penitentiary drove many into mental and physical destitution through solitary confinement, and plantations enslaved Black people who endured the degradation of being categorized as "things, not persons" (Stevenson, 2021, 279). In what ways is American history, and, in particular, the twin horrors of colonialism and slavery, represented in the U.S. ghost tourism industry? What kinds of stories are left in? And what is left out?

In *America's Horror Stories* we explore these questions by touring key sites of racist U.S. history. We begin in the coastal town of Salem, Massachusetts, where the 1692 Salem Witch Trials occurred. Two hundred were accused of witchcraft, including enslaved Arawak woman Tituba. We then travel to New Orleans, Louisiana. While popularly known for Mardi Gras, the city was also the largest slave trade hub in the American South. From here, we drive north-west to St. Francisville, Louisiana, which hosts the Myrtles Plantation, where enslaved Chloe is told to haunt the premises. We return north, to Gettysburg, Pennsylvania. Here, the bloodiest battle of the Civil War took place, as approximately 50,000 soldiers either died, were wounded, or went missing. Ghosts of Confederate and Union soldiers are told to continue haunting the land today.

Dark Tourism—Death and Politics

In Netflix's *Dark Tourist* (2018), New Zealand journalist David Farrier visits "unusual—and often macabre—tourism spots around the world." He meets a Jeffrey Dahmer enthusiast in Milwaukee, Wisconsin who takes him to locations popularized by the serial killer. In Los Angeles, California, Farrier meets friends and family of Charles Manson and tours spots where the cult figure directed his followers to commit murder. He visits Ōkuma, Fukushima, Japan, where residents evacuated due to the Fukushima Daiichi nuclear powerplant disaster on March 11, 2011. In Mexico City, he meets followers of Santa Muerte and witnesses an exorcism. From these examples, a definition of dark tourism comes to light. Dark tourism, broadly speaking, "is the act of travel to sites associated with death, suffering and the seemingly macabre" (Fonseca et al., 2016, 1).[1]

Dark tourism is not new. As Richard Sharpley (2009, 4–5) points out, "[V]isitors have long been attracted to places or events associated in one way or another with death, disaster and suffering," such as "the gladiatorial games of the Roman era, pilgrimages, and attendance at medieval public executions." Consider the Catacombs of Paris. The Catacombs were built in 1774, when basement walls collapsed around the Holy Innocents' Cemetery. The tunnels sit 20 meters underground and hold the remains of more than six million people. The underground cemetery has been featured on History Channel's *Cities of the Underworld* (2007) and Fox Family's *Scariest Places on Earth* (2009). Yet, over a century earlier, on April 2, 1897, roughly 50 amateur musicians played a night concert in the depths of the tunnels. In front of over 100 attendees, they performed a song inspired by the Catacombs, Saint-Saën's *Dance Macabre* (Olson, 2015). The *New Zealand Herald* (1897) described the event as, "A Gruesome Festival in Paris."

While off-putting for some, many are drawn to these sites for a number of reasons: to learn about the past, mourn atrocities, for morbid curiosity, voyeurism, *schadenfreude*, or just to have a thrilling experience (Sharpley, 2009, 17). Colin Dickey (2016, 213–215) describes thrill seeking by those searching for the ghost of Thelma Taylor, a woman who was abducted in 1949 under the St. Johns Bridge in Portland, Oregon. *Haunted Oregon* writer Andy Weeks (2014, 20) advises that readers "visit the area and see if you can hear the unearthly screams of now-deceased Thelma. They won't be pretty if you hear them, but ones that will make your skin crawl." As Dickey describes of the promotion, "What better way to spend a chilly evening than trying to scare yourself into feeling alive?" Corey Taylor (2014, 2) further explains his lifelong fascination with ghosts, stemming from childhood ghost hunting in Iowa, "[A]most everyone I know has a ghost story, and anyone who does not have one secretly and desperately wants one."

Dark tourism is not without issues. It can sensationalize and desensitize us to historical atrocities. John Lennon and Malcolm Foley (2000, 60–61) mention their discomfort when visiting Auschwitz, where 1.3 million were deported and 900,000 were killed. They wrote of "[o]bserving groups of casually dressed tourists taking snapshots, eating sandwiches, and spending only 90 minutes" at the concentration camp (paraphrased by Bowman and Pezzullo, 2009, 191). When visiting the Pennhurst Asylum, coauthor Favian Martín was astonished by the scale of commercialization on the site, as he examined numerous vendors selling paranormal trinkets and ghost hunting equipment, with a $40 fee to take a picture with the hosts of *Ghost Hunters* to boot. Indeed, the *New Zealand Herald* reported that the Catacombs of Paris concertgoers acted in a quite disrespectful manner,

> None of these mementoes of mortality seemed to trouble the midnight visitors to the catacombs. Some of them handled the skulls like Hamlet in the graveyard and indulged in mock moralisings over the dreamless head. A few cracked jokes on the ribs of death, and a medical professor volunteered a lecture on anatomy.

On the other hand, many visit sites of atrocity to explore their heritage and to confront the human capacity toward such violence (perhaps betraying the term "dark tourism" altogether). As Michael Bowman and Phaedra Pezzullo (2009, 189) remind us, "[T]ouring sites of or about the global slave trade, assassinations, and genocide further goads us to grapple with the human ability to act inhumanely and selfishly in seemingly unspeakable ways." Carolyn Ellis (Bochner and Ellis, 2016, 156) visited the memorial at the Treblinka concentration camp in Poland, where almost 80,000 died. She attended with 88-year-old survivor Jerry Rawicki and his three grandchildren and daughter-in-law. Ellis describes coming upon "the garden of approximately 17,000 multi-shaped and multi-colored stones ranging from the size of a hand to a large tombstone" (276). The stones symbolize matzevot—Jewish headstones (Muzeum Treblinka, n.d.). The group searched and found the Bodzentyn stone, named after the ghetto Jerry's mother and younger sister were taken (Bochner and Ellis, 2016, 277). Rawicki reflected on the visit, "Finally, I have been able to grieve. . . . Before I didn't have a place to come to, no cemetery, nothing" (278). Visitors also go to the Kilmainham Gaol in Dublin, Ireland, a prison which ran from 1796–1924 and was where 13 leaders of Irish independence were executed. In Stonebreakers Yard, a cross marks the spot leader James Connolly was killed by the British.

Whether one visits a site of atrocity for heritage, history, or thrill, all tourism is, to reiterate, historical—and is, therefore, political. As James Loewen (2018, 2) asserts, "Understanding our past is central to our ability to understand ourselves and the world around us." Given that dark tourism offers an avenue to understand the past—and the world around us—it provides, as Richard Sharpley (2009, 8) puts it, "[T]he opportunity to write or rewrite the history of people's lives and deaths, or to provide particular (political) interpretations of past events." Yet, these interpretations, he continues, can be highly "distorted, or displaced or disinherited" (12). What does dark tourism, and ghost tourism in particular, tell us about U.S. history—about America's horror stories of colonialism and slavery? To explore this, we must consider what exactly the "ghost" in ghost tourism is, and how the concept of the ghost, too, can help us understand the past, present, and future.

Social Violence and the Ghost(ed)

A Hollywood image often comes to mind when one thinks of a ghost. It may be Casper the Friendly Ghost, the Ghost Busters taking on the green Slimer, the demonic entity manipulating TV screen static in the *Poltergeist*, or the three hitchhiking ghosts populating Disneyland's Haunted Mansion ride. Yet, the concept of the ghost, or the *ghost(ed)*, has also been used by critical scholars as a metaphor to think about how past atrocities *haunt* today. As Hannah-Jones (2021, xxix) further states, "history molds, influences, and haunts us in the *present*." To this, Avery Gordon (2011, 2) defines haunting as "an animated state in which a repressed or unresolved social violence is making itself known, sometimes very directly, sometimes more obliquely."

Social violence is inscribed *within space* as a cultural, residual phenomenon (Ferrell, 2022, 75). In nonfiction novel *In Cold Blood*, Truman Capote (1966) tells of the 1959 murder of the Clutter family in Holcomb, Kansas which was perpetrated by Dick Hickock and Perry Smith. As Travis Linnemann (2015, 514) observes, the afterlife of the murders persists in true crime novels and crime reporting. The murders "haunt the minds of the public as the horrors of random crimes and senseless violence." In his nightly London excursions, Theo Kindynis (2019, 25) has searched for "years—or even decades—old graffiti." These markings are referred to in graffiti parlance as "ghosts." They represent, he discerns, "lingering and material and atmospheric traces of the past." Through them, Kindynis concludes, "Memory and trauma become inscribed literally, symbolically, affectively and atmospherically in space and place" (39). In *Ghostland*, Colin Dickey (2016, 255) also observes the physical traces of deindustrialized Detroit, Michigan. "In the slow decay" of the abandoned Roosevelt Warehouse, which turned into a book depository for the Detroit public schools,

> trees have sprouted from the wreckage and books and other supplies left behind, offering a particularly stark image of Detroit's abandonment and, to some extent, its rebirth—or at least its reclamation by nature.

What is remembered and how it is remembered is itself political. Is graffiti a representation of historical memory and culture, or is it vandalism? Are shuttered buildings a death of industry, a reclamation by nature, or both? On the politics of memory, Allison Young (2022, 229–234) explores collective memorialization for those who died in the 24-story Grenfell Tower. The complex was constructed between 1972 and 1974 as "social housing" in North Kensington, West London. On June 14, 2017, a fire started in a fourth-floor apartment, which was "the result of a faulty part in a refrigerator." Seventy-two individuals died over a 60-hour period. Residents established hundreds of memorials, with one reading in a heart on a brick wall, "There's no justice. It's just us." These markers acted, as Young reflects, as "spatial manifestations of bereavement, sadness, and anger," which "made visible the dreadful absence of those who had died and those who had lost their homes." Even after the memorials faded over time, or were "removed, cleaned, and stored safely" by a local cultural association, the haunting memory of the event lives on. Young concludes, "If we are to understand ghosts, our task is to find these origins, to listen to these echoes—to excavate absence for what remains present in it."

Indeed, dominant narratives often erase memories of such violence through what Warren Cariou (2006, 730) refers to as willed forgetfulness (Saleh-Hanna, 2015, para. 18).[2] Shanna Felix and Meredith Garcia (2023, 154–155) write about the haunting of Kitty Genovese, a queer white woman killed by Winston

Moseley in Manhattan, New York. Initial reporting focused on the "bystander effect" and did not mention her queer identity. Genovese's story, they emphasize, is therefore "important to the LGBTQ world because it speaks to a much larger horror experienced by the queer community"—that of victimization *and* its subsequent erasure. Yet, Genovese's whiteness, too, made her a sympathetic victim, with reporters disregarding others murdered by Moseley, including 24-year-old Mae Johnson, a Black woman. Consequently, Genovese's tailored visibility compounds the *invisibility* of Johnson. Kris Manjapra (2022, 4) describes such in/visibility through what he refers to as the *ghost line*,

> Societies draw veils to divide the realm of the seen and remembered from the realm of the systemically erased and disavowed. If the color line [of W.E.B. Du Bois (1903) fame] creates racial divides to oppress and dispossess, the ghost line creates existential divides between being and nothingness; between those said to be present and those designated as society's present-absences.

And, yet, Mae Johnson's murder, as well as the unaccounted murders of Black women more generally (see Combahee River Collective, 1979), still haunt/s Genovese's death and reporting thereafter. Thus, within present-absences there exists the ghostly figure of unresolved violence.

Searching for hidden ghosts is key to reconcile with such violence. In her 1987 novel *Beloved*, Toni Morrison (2019) tells of a ghost who haunts protagonist Sethe, Sethe's daughter, Denver, and Sethe's live-in partner, Paul D. The embodied spirit, named Beloved, is Sethe's two-year old daughter, of whom Sethe murdered to protect her from the horrors of slavery: "[I]f I hadn't killed her," Sethe contends of the white men who sought her capture, "she would have died and that is something I could not bear to happen to her" (200). Beloved visits the three in the figure of a 20-old woman. They must face her in their own way, and thus confront the collective, haunting presence of slavery and its aftermath. As Mohammad Deyab (2016, 13) offers, they must "discover and come to terms with themselves and their tasks within the African American community." Beloved's presence also allows Sethe's neighbors to recontextualize her act as, at once heinous, an act of desperation and love within the shared historical context of slavery (19).

Given this shared context, the ghost of Beloved reflects, as Deyab (2016) reviews, the *cultural haunting* of chattel slavery imposed on Black Americans. Indeed, Morrison relied on historical evidence to craft the novel. First, she drew on newspaper clippings of Margaret Garner, an escaped enslaved person who, like Sethe, killed her daughter to protect her from the indignity of slavery. Garner was, the *American Baptist* (Bassett, 1856) reported, "unwilling to have her children suffer as she had done" (Singh, 2021). The second form of historical evidence is "the horrible history of Middle Passage where millions of African Americans lost their souls in their journey." For Morrison, as Deyab (2016, 17) paraphrases, "each of these dead souls has a ghost that wants to be remembered and accounted for; and therefore, wants to bridge the gap between what is past and what is present."

Inspired by such inquiry, Viviane Saleh-Hanna (2015, para. 20) recommends a Black feminist hauntology to capture "the expanding and repetitive nature of structural violence" that continues to haunt present institutions, as white settlements have turned into urban centers, plantations into prisons, slave catchers into police, and images of the rebellious Black slave into the "dangerous criminal" (para. 49; Baucom, 2001, 80).[3] As Saleh-Hanna (2015, para. 32) states aptly, "Ghosts remain because justice has not been achieved"—so we must discover the present-absences of ghosts to reconcile with the past. For Ta-Nehisi Coates (2015, 12), the constant interrogation of racism's haunting of past and present affords personal and collective liberating potential, "The greatest reward of this constant interrogation, of confrontation with the brutality of my country, is that it has freed me from ghosts and girded me against the sheer terror of disembodiment."

Not merely past and present social violence, but the ghost(ed) also forecasts future violence. As Michael Fiddler and colleagues (2024, 1) contend, "[T]he present is haunted by the future—either our visions for an imagined future or '[f]uture injuries that are felt in the now'" (paraphrased by Petersen, 2024, 6). Amanda Petersen (6) observes the haunting future of police violence in award winning photos submitted to the federally funded Office of Community Oriented Policing Services (COPs). The program encourages non-federal police agencies to submit one photo per year that best reflects community engagement. Many of them depict police kneeling in front of Black children, with their gun visible in holster. Petersen (10–11) cites such imagery as portending—and obscuring—future racist state violence:

> Even without the dominance of the gun, the pattern of kneeling itself conjures a future haunting of [Minneapolis police officer] Derek Chauvin lethally kneeling on the neck of George Floyd in 2020, years after several of these winning photos were taken.

In turn, unresolved social violence—as felt through the ghosts of the past—does much to haunt our *capacity* for future social justice (see Fisher, 2014).

It warrants a reminder that the scholars cited here do not speak of "ghosts or the supernatural" in the "real" sense of the terms. Rather, they utilize ghosts, or the ghost(ed), as "conceptual metaphors, allowing us to *see* what we *feel* haunting us" (Fiddler et al., 2022, 3–5). Ghosts are returned spirits (Miles, 2015, 12) who "announce the unended," as Manjapra (2022, 9–10) contends when discussing the present-day hauntings of chattel slavery. Ghosts "trouble postslavery societies in our consciences, our memories, and our social disorders, in order to disturb us and ask us for redress." The concept of the ghost, then, calls us to bear witness to those who have been subjected to structural violence(s)—and of whom have faced premature death (Gilmore, 2007, 28).[4] The ghost furthermore calls us to observe "how the harms that they have suffered have been obscured or redacted" and how this redaction bears on present and future injustice (Fiddler at al., 2022, 5).

While we use ghosts as a metaphor for the haunting of unresolved social violence, ghosts marketed within the dark tourism industry are nevertheless taken quite literally: ghosts appear by tour guides in photos, sounds, and spine-tingling sensations. While we do not here debate whether ghosts exist as a supernatural phenomenon, we do consider how ghosts are *socially made* by ghost tour guides, particularly on land haunted by past injustices—and how the social creation of ghosts may work to obscure—or reveal—haunting past-present-future violence.

Ghost Tourism and Spectral Sensation-Making

Within a sociological context ghosts are made through social *performance*. Much like the stage performer, as Erving Goffman (1959) highlights, in social interaction participants *perform* to each other by acting out socially validated "scripts" and "roles." The performance occurs within a social setting, or on a "stage." Restaurants, for instance, provide a "backstage" kitchen where waiting staff can talk about customers in a way that would not be socially acceptable in the "frontstage" dining area. The same can be said for tourism, where the destination is a "stage" and tour guides and visitors both act out their "roles" through proper "scripts." While guides are expected to offer the history of any given place, following expected social scripts and at times actual scripts provided by the tour company, visitors are certainly not expected to do so—and if they try to, tension will arise. When each "actor" plays their role, the shared performance creates a collectively defined social reality. As Jane Desmond (2004) explains, "When tourists travel . . . what they witness isn't merely *like* a performance, it *is* a performance insofar as the site is often composed of live bodies engaged in acts that are put on display for tourists" (paraphrased by Bowman and Pezzullo, 2009, 193; MacCannell, 2013).

Within the ghost tourism industry, ghosts are created—or *performed*—by guides in numerous ways. As guides take travelers through cemeteries, alleyways, or orphanages, they share "spirit" photos of ghostly orbs, offer stories of ghostly encounters, and encourage participants to say something if they see something. They seek to make participants *feel* the presence of ghosts by signaling to "shadow figures, phantom odors, and voice phenomena" (Riley, 2014, 4). Ghost tour guides are, thus, in the work of *sensation-making*:[5] they "say and do things" (Loseke, 2017, 21) in order to foster a haunting atmosphere and affect conducive for ghost sightings (Young, 2014).

Certainly, ghost performances are shaped by the type of tour provided, or the "stage" which is set. In walking tours, visitors alternate paces, meander, and look between buildings, affording a "more complete sense of place" (Gentry, 2007, 225; Tuan, 1977). They talk to guides, engage with pedestrians, and navigate traffic, weather, and property. Site-based tours, such as those in prisons and plantations, permit a similar kind of exploration, although it is limited to accessible spots on the property. Driving tours cover more ground, protect participants from direct weather exposure, and may be more accessible for those using crutches or a wheelchair. Yet, they also "limit[] the tourists' gaze to street level" (Gentry, 2007, 231). Self-guided tours remove direct interaction with guides with smartphone-based applications offering audio recordings or textual descriptions do the work. Self-guided tours offer flexibility in terms of where one goes and for how long. Regardless of the approach, each provides a medium for affective performance—each is a stage to which the ghosted social performance plays out.

Ghost hunting tools aid in the social creation of ghosts, whether they are provided by the tour company or brought by participants. Tourists may seek spirit voices through radio frequency recorded on electronic voice phenomenon (EVP) or through an electric field meter (EMF) to monitor changes in the electromagnetic environment which may indicate ghostly presences. Others hold L-shaped copper dowsing rods. If there is a presence, the rods move synchronously. The Ghost Radar phone application also utilizes a radar and camera scanner to detect paranormal activity. One does not need advanced ghost technology, however, the camera has long been used to capture spectral photos of white, round spirit orbs or humanlike shapes (Manseau, 2017). To be sure, mass ghost media is also incorporated into the performance, as tourist sites, books, and the abundance of ghost television series like *Ghost Hunters* shape the expectations of visitors to see ghosts and inform how the technology works.

Ghost hunting technologies are subsequently integrated into the social performance. After Dickey (2016, 90) detected a ghostly figure on his camera during a hunt, his friend admitted, "[I]f you toggle your infrared lights on and off while standing near someone else, the interference will cause orbs and shadows to appear on the person's video." Dickey concluded from this experience, "You can, in other words, create your own ghosts." Christopher Bader and colleagues (2017, 98) have found from their excursions that ghost hunting technology creates ambiguous sensory input that "blurs perception, substituting the uncanny for the everyday and firing the imagination." The fired-up imagination is given further credence by fellow seekers yearning for a ghostly experience (102). To use another metaphor, ghost hunters adopt a ghost hunting habitus: they are expected to utilize ways of seeing, acting, and being to which ghosts are socially conjured (Bourdieu, 2020).

Performances can also be disrupted (Goffman, 1959, 14). Simply, tourists do not need to play along with the guided experience. When touring the UK prison museum Galleries of Justice in Nottingham, Diane Urquhart wrote of feeling pressure to witness spiritual activity. She did not offer the anticipated reaction,

> Our guide asks for a volunteer and somehow I am ushered to a chair directly facing the mirror. I am given a small torch and am instructed to place it upright under my chin. A mixture of light and shade is cast eerily across my face. As I look into the mirror, my peers are invited to look with

me, and we wait. "What can you see?" asks our guide. We all stare, perhaps a little uncomfortably, and without reply. The position of the torchlight is adjusted by the guide: "I can see the eyes sinking inwards, they look darker . . .", he continues. At this, there is a gentle nod from an onlooker, ". . . and can anyone else see the outline of a beard?"

I cannot resist the urge to crack a joke at this point and the group laughs quietly, but their squinting tilted heads remain fixed on my reflection; their desire to witness some form of transformation abundantly clear.

(Hodgkinson et al., 2017, 572–573).

In this regard, the visitor is not a passive spectator but may be described as a "spect-actor" (Boal, 1985). As Michael Bowman and Phaedra Pezzullo (2009, 193) remind us, "[T]ourists are not so much 'consumers' of 'products' as they are audiences of multimedia, oral-dramatic events."

To be sure, America's horror stories—such as that of colonialism and slavery—*haunt* the land ghost tours are on. Plantations are reused, rebuilt, and remembered, the aftermath of slavery acting as residual phenomenon of which are obscured, redacted, or engaged with during ghost tours (Ferrell, 2022, 75). And, while some ghosts may be considered to haunt a space by guides in ways that obscure the past, ghosts also open an avenue to consider historical violence, and to begin to reckon with it.

Ghost Tour(ist) Methodology and Book Outline

With these conceptual tools, we embark in this book to Salem, New Orleans, St. Francisville, and Gettysburg. To conduct research, we went on walking, driving, and self-guided tours. We stayed in "haunted" hotels, browsed local shops, and dined at restaurants. We analyzed social and mass media, snapped pictures, and kept field notes. We visited, toured, and returned home, making our approach a uniquely ghost tour(ist) methodology. Indeed, we would come to find that the tourist experience is quite ghostlike: of being caught in a kind of purgatory, or liminal space, eating the same breakfast at a hotel, moving along the same shops, and seeing tourists coming and going as ghostly hordes (McKay, 2022).[6] From this, we identified several themes.

From our travels we would come to find that, first and foremost, the ghost tourism industry largely whitewashes past atrocities by ignoring or downplaying the impact of colonialism and slavery. Guides at Gettysburg, for instance, made no mention of slavery's role on Confederate secession, and those in Salem did little to discuss the role of slavery in the Massachusetts Bay Colony. Second, ghost tour stories in the industry tend rely on racist stereotypes. Enslaved Chloe, of whom is told to haunt the Myrtles Plantation, was represented through a variety of controlling images of Black women, such as the Jezebel and mammie tropes (Collins, 2008). Stories of Madame Delphine LaLaurie's brutal treatment of those she enslaved in New Orleans presented them as a faceless animalistic mass (Foley, 2021, 68). Third, ghost stories represent punishment in a way that is fundamentally voyeuristic and detached, inhibiting a deeper interrogation of how state violence operated then and continues to operate today (Morris and Arford, 2019, 432; Brown, 2009, 12–13). Fourth, the ghost tourism market profits by selling bits and pieces of spiritual belief, such as Voodoo spells for success in relationships and finances. With little engagement of historical atrocity, guides ultimately appeal to a sensitive (white) consumer who has little to no interest in engaging with historical atrocity and their place within it.

While the ghost tourism industry has obscured the past, as we argue here, the *ghost(ed)* also offers an avenue to reveal how past atrocities haunt the visited sites, whether guides recognize it or not. As such, we consider a hauntologically informed examination of space (Fiddler, 2019, 474) to consider a larger scope of historical atrocity and resistance. We reimagine figures like Chloe at the Myrtles Plantation or enslaved Tituba in Salem as representing invisibilized Black and Indigenous oppression, agency, and resistance. When Tituba falsely confessed to witchcraft, she utilized a variety of African, Indigenous, and English techniques to avoid further persecution (Breslaw, 1995). Chloe, too, can be understood as representing the Black feminist sociological ghost who depicts the atrocity of and resistance toward chattel slavery (Saleh-Hanna, 2015, para. 20). She reveals the "unacknowledged ghosts" (Batchen, 2008, 10) of those enslaved largely not discussed by guides.

We do not discount dark tourism in its entirety. Certainly, ghost tourism—and the larger dark tourism industry—is not going away anytime soon. Rather, as we discuss in the final chapter, we recommend a critical ghost tourism that considers the unresolved social violence that haunts us today: of engaging tours to see what is told, how it is told, and what is left out; then of returning to that of which is left out to reconjure obscured past violence and struggle in what bell hooks (2012) comprehensively refers to as the imperialist white supremacist capitalist patriarchy. This requires seeing, in sites of tourism, real American horrors: of colonialism, slavery (Wedderburn, 1824), and lynching (Wells, 1892).[7] After all, as Kelly Hayes and Mariame Kaba (2023, 103) put it, the character of and connections between the "unspeakable horrors that occurred on unthinkable scales" through genocide and chattel slavery, "frame every horror that the United States has perpetrated since."

To elucidate rather than obscure these horrors, we recommend a critical ghost tourism that centers social justice (Ateljevic et al., 2007). This requires an ethics of respect and reflexivity (Dalton, 2014, i), including that of the privileged status of the tourist in general, as well as the various social statuses they may occupy. To this, author Kevin is white and author Favian is Latino, and we both reside on the stolen land of the Lenape People in what is now Pennsylvania. Throughout the text we also discuss structural violence meted out on Black and Indigenous People, and we hope to do so with the kind of respect and reflexivity that we espouse as key for a social-justice minded ghost tourism. In adopting such critical stances, we take great inspiration from Black historian Tiya Miles (2015, 132) who wrote in *Tales from the Haunted South*,

> We *can* call forth the power of ghost as scholars, writers, artists, teachers, and stewards of historic sites, as indeed we must if we are to place progressive social justice visions in contention with a culture possessed by ghost fancy. But let our ghosts be real, let our ghosts be true, let our ghosts carry on the integrity of our ancestors.

Conclusion—Staying with Ghost Stories

In *Staying with the Trouble*, Karla Haraway (2016, 12) reminds us, "It matters what stories make worlds, what worlds make stories." In adapting her words, we might say: it matters what ghost stories make worlds, what worlds make ghost stories. Within the ghost tourism industry ghost stories tend to be sensationalized, commodified, and whitewashed. Nevertheless, ghost stories also drive critical histories for engagement with what haunts the past-present-future. Through ghost stories, we can unsilence the past by seeking the ghostly whispers of those subjected to structural violence. As Michel-Rolph Trouillot (2015, xix) contends, "history is the fruit of power," and "the ultimate mark of power may be its invisibility; the ultimate challenge, the exposition of its roots" (Hannah-Jones, 2021, xxx). If we may mix metaphors, staying with ghost stories, as with trouble, seeks to expose these roots.

To this, we do not ask the reader to be skeptical of ghosts as a supernatural phenomenon. That is up to them. We do, however, challenge a general skepticism in a "colorblind," postracial society of the ongoing legacy of colonialism, slavery, and genocide and to stay with ghost stories to reckon with past unresolved social violence. Only when we acknowledge America's horror stories can the haunting past truly be exorcised.

Notes

1 Dark tourism has also been referred to as thanatourism (Seaton, 1996, 131), morbid tourism (Blom, 2000), black spot tourism (Rojek, 1993), and fright tourism (Sharpley, 2009, 10).
2 Viviane Saleh-Hanna (2015, para. 18) defines willed forgetfullness as "a process whereby the colossal injustice in the settlement of the West has been successfully repressed by settler culture."
3 This develops Jacques Derrida's (1994) hauntology, of which "replac[es] the priority of being and presence with the figure of the ghost as that which is neither present nor absent, neither dead nor alive" (Davis, 2005, 373).
4 Ruth Gilmore (2007, 28) defines racism as "the state-sanctioned or extralegal production and exploitation of group-differentiated vulnerability to premature death."
5 Our description of guides as "sensation-makers" extends social constructionists' use of the term "claims-makers" in social problem studies (Loseke, 2017, 21)
6 We went to Gettysburg in June 2022 (nine tours), New Orleans (ten tours) and St. Francisville (three tours) in May 2023, Salem in August 2023 (ten tours), and we

revisited Gettysburg in June 2024 (three tours). In addition, we went to museums, shops, and memorial sites and did some ghost hunting of our own.
7 To be sure, the horrors of white supremacy are inverted through a cultural representation (see Hall, 1997) of marginalized groups as being monstrous—and a threat to white social order. As Frantz Fanon ([1952] 2008, 124–125) observes of the villainous Black and Indigenous person in the media:

> And the Wolf, the Devil, the Wicked Genie, Evil, and the Savage are always represented by Blacks or Indians; and since one always identifies with the good guys, the little black child, just like the little white child, becomes an explorer, an adventurer, and a missionary "who is in danger of being eaten by the wicked Negroes."

References

Ateljevic, Irena, Nigel Morgan, and Annette Pritchard. 2007. "Editors' Introduction: Promoting an Academy of Hope in Tourism Enquiry." In *The Critical Turn in Tourism Studies: Innovative Research Methodologies*, edited by Irena Ateljevic, Annette Pritchard, and Nigel Morgan, 1–8. Oxford: Elsevier.

Bader, Christopher D., Joseph O. Baker, and F. Carson Mencken. 2017. *Paranormal America: Ghost Encounters, UFO Sightings, Bigfoot Hunts, and Other Curiosities in Religion and Culture* (2nd ed.). New York: New York University Press.

Bassett, P.S.1856. "A Visit to the Slave Mother Who Killed Her Child." *American Baptist.*

Batchen, Geoffrey. 2008. "Snapshots: Art History and the Ethnographic Turn." *Photographies*, 1(2): 121–142.

Baucom, Ian. 2001. "Specters of the Atlantic." *South Atlantic Quarterly*, 100(1): 61–82.

Becker, Elizabeth. 2013. *Overbooked: The Exploding Business of Travel and Tourism.* New York:Simon & Schuster.

Blom, Tom. 2000. "Morbid Tourism—a Postmodern Market Niche with an Example from Althorp." *Norwegian Journal of Geography*, 54(1): 29–36.

Boal, Augusto. 1985. *Theater of the Oppressed* (trans. C. A. McBride and M.-O. L. McBride). New York: Theatre Communications Group.

Bochner, Arthur, and Carolyn Ellis. 2016. *Evocative Autoethnography: Writing Lives and Telling Stories*. New York and London: Routledge.

Bourdieu, Pierre. 2020. "Outline of a Theory of Practice." In *The New Social Theory Reader*, edited by Jeffrey C. Alexander and Steven Seidman, 80–86. New York and London: Routledge.

Bowman, Michael S., and Phaedra C. Pezzullo. 2009. "What's So 'Dark' about 'Dark Tourism'?: Death, Tours, and Performance." *Tourist Studies*, 9(3): 187–202.

Breslaw, Elaine G. 1995. *Tituba, Reluctant Witch of Salem: Devilish Indians and Puritan Fantasies*. New York: New York University Press.

Brown, Michelle. 2009. *The Culture of Punishment: Prison, Society, and Spectacle*. New York: New York University Press.

Capote, Truman. 1966. *In Cold Blood*. New York: Random House.

Cariou, Warren. 2006. "Haunted Prairie: Aboriginal 'Ghosts' and the Specters of Settlement." *University of Toronto Quarterly*, 75(2): 727–734.

Coates, Ta-Nehisi. 2015. *Between the World and Me*. Melbourne: Text Publishing.

Collins, Patricia Hill. 2008 [1990]. *Black Feminist Thought: Knowledge, Consciousness, and the Politics of Empowerment*. New York and London: Routledge.

Combahee River Collective. 1979. "Eleven Black Women: Why Did They Die?" *Third World Women.*

Dalton, Derek. 2014. *Dark Tourism and Crime*. New York and London: Routledge.

Dark Tourist. 2018. Netflix.

Davis, Colin. 2005. "Hauntology, Spectres and Phantoms." *French Studies*, 59(3): 373–379.

Derrida, Jacques. 1994. *Specters of Marx: The State of Debt, the Work of Mourning, and the New International*. New York and London: Routledge.

Desmond, Jane C.2004. *Staging Tourism: Bodies on Display from Waikiki to Sea World*. Chicago: University of Chicago Press.

Deyab, Mohammad Shaaban Ahmad. 2016. "Cultural Hauntings in Toni Morrison's Beloved (1987)." *English Language, Literature & Culture*, 1(3): 13–20.

Dickey, Colin. 2016. *Ghostland: An American History in Haunted Places*. Westminster: Penguin.

Du Bois, W.E.B.1903. *The Souls of Black Folk*. Chicago: McClurg.

Fanon, Frantz. 1952 [2008]. *Black Skin, White Masks*. New York: Grove Press.

Felix, Shanna N., and Merideth Garcia. 2023. "The Hauntings of Kitty Genovese: The Bystander Effect and Queer Invisibility." In *The (Mis) Representation of Queer Lives in True Crime*, edited by Abbie E. Goldberg, Danielle C. Slakoff, and Carrie L. Buist, 141–159. New York and London: Routledge.

Ferrell, Jeff. 2022. "Ghost Method." In *Ghost Criminology: The Afterlife of Crime and Punishment*, edited by Michael Fiddler, Theo Kindynis, and Travis Linnemann, 67–87. New York: New York University Press.

Fiddler, Michael. 2019. "Ghosts of Other Stories: A Synthesis of Hauntology, Crime and Space." *Crime, Media, Culture*, 15(3): 463–477.

Fiddler, Michael, Theo Kindynis, and Travis Linnemann. 2022. "Ghost Criminology: A (Spirit) Guide." In *Ghost Criminology: The Afterlife of Crime and Punishment*, edited by Michael Fiddler, Theo Kindynis, and Travis Linnemann, 1–31. New York: New York University Press.

Fiddler, Michael, Linnemann, Travis, and Theo Kindynis. 2024. "Ghost Criminology: A Framework for the Discipline's Spectral Turn." *British Journal of Criminology*, 64(1): 1–16.

Fisher, Mark. 2014. *Ghost of My Life: Writings on Depression, Hauntology, and Lost Futures*. Alresford: Zero Books.

Foley, Laura. 2021. "The Haunted History of New Orleans: An Exploration of the Intersectionality between Dark Tourism, Black History, and Public History." *ProQuest Dissertation Publishing*. Glassboro: Rowan University.

Fonseca, Ana Paula, Claudia Seabra, and Carla Silva. 2016. "Dark Tourism: Concepts, Typologies and Sites." *Journal of Tourism Research & Hospitality*. http://dx.doi.org/10.4172/2324-8807.S2-002.

Gentry, Glenn W. 2007. "Walking with the Dead: The Place of Ghost Walk Tourism in Savannah, Georgia." *Southeastern Geographer*, 47(2): 222–238.

Ghost City Tours. n.d. "The Ghosts of the Lyceum and Turner's Seafood."https://ghost citytours.com.

Gilmore, Ruth Wilson. 2007. *Golden Gulag: Prisons, Surplus, Crisis, and Opposition in Globalizing California*. Berkeley: University of California Press.

Goffman, Erving. 1959. *The Presentation of Self in Everyday Life*. New York: Knopf Doubleday Publishing Group.

Good Morning America. 2017. "Ghost Selfie Going Viral."

Gordon, Avery. 2011. "Some Thoughts on Haunting and Futurity."*Borderlands*, 10(2): 1–21.

Hall, Stuart. 1997. *Representation: Cultural Representations and Signifying Practices*. London: Open University.

Hannah-Jones, Nikole. 2021. *The 1619 Project: A New Origin Story*. New York: One World

Haraway, Donna J. 2016. *Staying with the Trouble: Making Kin in the Chthulucene*. Durham: Duke University Press.

Hartman, Saidiya. 2022. *Scenes of Subjection: Terror, Slavery, and Self-Making in Nineteenth-Century America*. New York: W.W. Norton & Company.

Hayes, Kelly, and Mariame Kaba. 2023. *Let This Radicalize You: Organizing and the Revolution of Reciprocal Care*. Chicago: Haymarket Books.

Hodgkinson, Sarah, and Diane Urquhart. 2017. "Ghost Hunting in Prison: Contemplating Death through Sites of Incarceration and the Commodification of the Penal Past." In *The Palgrave Handbook of Prison Tourism*, edited by Jacqueline Wilson, Sarah Hodgkinson, Justin Piché, and Kevin Walby, 559–582. London: Palgrave Macmillan.

hooks, bell. 2012. *Writing beyond Race*. New York and London: Routledge.

Kindynis, Theo. 2019. "*Excavating Ghosts: Urban Exploration as Graffiti Archaeology*." *Crime, Media, Culture*, 15(1): 25–45.

Linnemann, Travis. 2015. "Capote's Ghosts: Violence, Media and the Spectre of Suspicion." *British Journal of Criminology*, 55(3), 514–533.

Linnemann, Travis. 2022. *The Horror of Police*. Minneapolis: University of Minnesota Press.

Loewen, James W. 2008. *Lies My Teacher Told Me: Everything Your American History Textbook Got Wrong*. New York: The New Press.

Loseke, Donileen. 2017. *Thinking about Social Problems: An Introduction to Constructionist Perspectives*. New York and London: Routledge.

Luhrmann, Tina. 2013. "Conjuring up Our Own Gods." *New York Times*. https://www.nytimes.com.

MacCannell, Dean. 2013. *The Tourist: A New Theory of the Leisure Class*. Berkeley: University of California Press.

Manjapra, Kris. 2022. *Black Ghost of Empire: The Long Death of Slavery and the Failure of Emancipation*. New York: Simon & Schuster.

Manseau, Peter. 2017. *The Apparitionists: A Tale of Phantoms, Fraud, Photography, and the Man Who Captured Lincoln's Ghost*. Boston: Houghton Mifflin Harcourt.

McKay, Carolyn. 2022. "Who's Been Sleeping in My Bed? Cheap Motel Rooms and Transgression." In *Ghost Criminology: The Afterlife of Crime and Punishment*, edited by Michael Fiddler, Theo Kindynis, and Travis Linnemann, 280–306. New York: New York University Press.

Miles, Tiya. 2015. *Tales from the Haunted South: Dark Tourism and Memories of Slavery from the Civil War Era*. Chapel Hill: University of North Carolina Press.

Morris, Patricia, and Tammi Arford. 2019. "'Sweat a Little Water, Sweat a Little Blood': A Spectacle of Convict Labor at an American Amusement Park." *Crime, Media, Culture*, 15(3): 423–446.

Morrison, Toni. 2019. *Beloved*. New York: Vintage Books.

Muzeum Treblinka. n.d. "Commemoration." https://muzeumtreblinka.eu.

New Zealand Herald. 1897, June 5. "A Gruesome Festival in Paris." xxxiv.10461.

Olson, Breanna. 2015, June 19. "Music in the Catacombs." *Musical Geography*.

Petersen, Amanda M. 2024. "Community-Oriented Copaganda: Anti-Black Violence in a Visual Archive of Policing." *Crime, Media, Culture*. doi:17416590241231904.

Riley, Johlene. 2014. *Ghost Hunting-The Gettysburg Files*. Arbor House at Gettysburg

Rojek, Chris. 1993. *Ways of Escape: Modern Transformations in Leisure and Travel*. London: Palgrave MacMillian.

Saleh-Hanna, Viviane. 2015. "Black Feminist Hauntology. Rememory the Ghosts of Abolition?" *Champ Pénal/Penal Field*, 12.

Scariest Places on Earth. 2009. *Cities of the Underworld*.

Seaton, Tony. 1996. "From Thanatopsis to Thanatourism: Guided by the Dark." *International Journal of Heritage Studies*, 2:234–244.

Sharpley, Richard. 2009. "Shedding Light on Dark Tourism: An Introduction." In *The Darker Side of Travel: The Theory and Practice of Dark Tourism*, edited by Richard Sharpley and Philip R. Stone, 3–22. Bristol: Channel View Publications.

Singh, Amardeep. 2021. *The Story of Margaret Garner: Inspiration for 'Beloved'*. Lehigh University.

Stevenson, Bryan. 2021. "Punishment." In *The 1619 Project: A New Origin Story*, edited by Nikole Hannah-Jones, 275–283. New York: One World.

Stone, Philip R. 2011. "Dark Tourism Experiences: Mediating between Life and Death." In *Tourist Experience: Contemporary Perspectives*, edited by Richard Sharpley and Philip R. Stone, 21–27. New York and London: Routledge.

Taylor, Corey. 2014. *A Funny Thing Happened on the Way to Heaven (Or, How I Made Peace with the Paranormal and the Stigmatized Zealots and Cynics in the Process)*. Cambridge: Da Capo Press.

Trouillot, Michel-Rolph. 2015. *Silencing the Past: Power and the Production of History*. Boston: Beacon Press.

Tuan, Yi-Fu. 1977. *Space and Place: The Perspective of Experience*. Minneapolis: University of Minnesota Press.

WDSU. 2021, February 26. "Do You See It? Louisiana Plantation Posts Viral Photo of Supernatural Sighting." https://www.wdsu.com.

Wedderburn, Robert. 1824. *"The Horrors of Slavery."*

Weeks, Andy. 2014. *Haunted Oregon: Ghosts and Strange Phenomena of the Beaver State*. Mechanicsburg: Stockpole Books.

Wells, Ida B. 1892. *Southern Horrors: Lynch Law in All Its Phases*.

Young, Alison. 2014. "From Object to Encounter: Aesthetic Politics and Visual Criminology." *Theoretical Criminology*, 18(2): 159–175.

Young, Alison. 2022. "9. The Time of Ghosts: Sites of Violence, eEnvironments of Memory." In *Ghost Criminology: The Afterlife of Crime and Punishmen*, edited by Michael Fiddler, Theo Kindynis, and Travis Linnemann, 227–252. New York: New York University Press.

Yuko, Elizabeth. 2021, October 28. "The Terrifying Rise of Haunted Tourism." Bloomberg. https://www.bloomberg.com.

1

SUMMONED TO SALEM

Introduction

English Puritans settled in the Massachusetts Bay Colony in the late 1620s. Concerned that the Church of England was reverting to its Catholic ways, they sought refuge in the so-called New World. The land was divine providence, a "new Jerusalem," where settlers could interpret scripture without the authority of a monarch or a Pope. Between 1630 and 1643, 21,200 Puritans fled England on 198 ships. One area they arrived in was Salem. Roger Conant settled the territory in 1628, which had been occupied by the Naumkeag band of the Massachusetts Tribe. Within a year, two hundred Puritans arrived. By 1637, the number increased to 1,000. Spurred by such growth, Salem divided into Salem Town (present-day Salem City) and Salem Village (present-day Danvers). The latter was a farming village and was where Samual Parris was appointed as Salem minister by Reverend Cotton Mather in July 1689. It was under Parris's roof where the event occurred that sparked the Salem witch trials (see Gagnon, 2021).

In January 1692, Parris's nine-year-old daughter Betty and 11-year-old niece Abigail Williams[1] "suddenly became ill and had seizure-like fits" (Roach, 2004, 7). They "climbed into holes, crawled under furniture, and struck 'sundry odd postures and antic gestures, uttering foolish, ridiculous speeches, which neither they themselves nor any others could make sense of'" (Lawson, 2002, 162). Parris sought to handle the fits in secret. He hired several doctors privately, but they did not find a medical explanation. So, he turned to local physician William Griggs. Dr. Griggs concluded the worst: the girls had been touched by an "evil hand"; they were bewitched. The next day, when Parris was out of town attending a sermon, his neighbor Mary Sibley recommended that Tituba and John Indian, who were both enslaved by Parris, bake a witch's cake. Per

DOI: 10.4324/9781003397809-2

English folk magic, the cake contained the girls' urine and was fed to the family dog. It was expected that the dog would reveal the names of those bewitching the girls. For dabbling in such magic, Parris whipped Tituba and questioned the girls' activities. They provided the names of three women as the source of witchcraft: Sarah Good, Sarah Osborne, and Tituba (Gagnon, 2021, 79–80).

On March 1, the three condemned women were brought into questioning by magistrates Jonathan Crown and John Hathorne. Good and Osborne maintained their innocence. Yet, like so many accused of witchcraft, Tituba falsely confessed. She spoke of sabbaths, riding on broomsticks, and that she, as with Good and Osborne, signed the Devil's Book. She added four names to the book and described "a tall man of Boston. . . . He goes in Black clothes, a tall man with white hair." Accusations began to spread, both by children and adults. The guilt of those accused was determined with spectral evidence, that of "the accusers' claims of seeing and being harmed by specters that were invisible to everyone else," which certainly made it difficult for those accused to mount a defense (Gagnon, 2021, 164). Between February 1692 and May 1693, more than two hundred were accused of witchcraft, including Bridget Bishop, Rebecca Nurse, and Martha Corey. Nineteen were hanged, four died in jail, and one, Giles Corey, was pressed to death.

Multiple contextual factors have been offered to make sense of the trials (see Gagnon, 2021, 63–66).[2] The Puritans made the grueling travel across the Atlantic Ocean. Upon arrival, many contracted smallpox, diphtheria, and malaria (Breslaw, 1995, 85). They were at war with French Catholics from Ottawa and Indigenous people defending their land. They were also at odds with the Church of England. Indeed, Massachusetts lost its Royal Charter between 1689 and June 1692, putting the colony in political limbo. There were land disputes, food scarcity, and drought, and Salem Village had rotating ministers. Parris furthermore received local criticism for gaining possession of the ministers' parsonage. He also spent hours speaking of evil in his sermons. More generally, though, the trials offered proof that the Devil existed and could work through community members, keeping Puritan followers within moral order.

To be sure, the witch hunts already had a long history in Europe before the Salem Witch Trials. Between 1300 and 1850, 42,215 were tried and 16,152 were executed across Europe, with the most in Germany (16,474 tried and 6,887 executed; McCarthy, 2019). The origins of the hunts occurred when Christian clergymen, referred to as demonologists, associated magic with devil worship. Previously magic was not considered to be inherently good or evil, and many had used magic "to heal, divine, remove effects of witchcraft, trace lost or stolen goods, or induce one person to love another" (Hutton, 2017, 78). Demonologists, by contrast, contended that *all magic* was the work of the devil, with witches having consigned themselves to demonic worship. While not exclusively so, women were the primary targets. Monk and inquisitor Heinrich Kramer argued in his 1484 Malleus Maleficarum (Hammer of the Witches) that women were especially prone to devil worship, because they were inherently "wicked" and "more credulous" than men. "[S]ince the chief aim of the devil is to corrupt faith, therefore he attacks them [women]," Kramer claimed (17–18; see Gibson, 2024). The witch hunts also disciplined women into unpaid household labor within the expanding capitalist order, as they previously acted "as healers, folk doctors, herbalists, midwives, [and] makers of love-philters" (Federici, 2004, 27). As Mona Chollett (2022, 211) puts it, "disorderly women, like chaotic nature needed to be controlled," and the witch hunts were one way of doing so (see also Ben-Yehuda, 1980).

With local and international witch hunts largely considered a past atrocity, Salem now markets itself as a tourist destination through a more playful image of the witch. Since the 1970s, tourism has become the most profitable industry in Salem, outcompeting industrial, mercantile, and fishing businesses (DeRosa, 2009, 157)—and the broomstick-flying witch has been a key part of this growth. The city brands itself as "Witch City" and is considered, by one tourist website, as "the most haunted place in America" (Meisler, 2024). Ghost tour companies abound, including the "Haunted Footsteps Ghost & Paranormal Salem Tour," "Salem Ghosts: Witches, Warlocks, & Hauntings Tour," and "Salem's Spooky Spectres Walking Ghost Tour." Tour guides take visitors to trial-related locations, including cemeteries, houses, taverns, hotels, and restaurants (Guiley, 2011; Baltrusis, 2014). The industry, too, is bolstered by an abundance of media attention, ranging from books like *House of the Seven Gables* (1851),[3] plays like *The Crucible* (1953), movies like Disney's *Hocus Pocus* (1993), and television series like *American Horror Story: Coven* (2013–2014).

How does the dark tourism industry in Salem represent the American horrors of the New England past—of colonialism, slavery, and the witch hunts? As we discuss in this chapter, the dark tourism industry in Salem has continued to rely on the stereotypical "Hollywood" image of the broomstick-flying witch, sensationalizing the trials in marketable fashion. In taking visitors to places of punishment, tour guides also frame the trials as representing past state violence isolated from today. By consigning the focus of tours to the trials, guides also do not discuss the impact of colonialism, slavery, and racist ideas in the region. Yet, as we add, the story of Tituba (or "Tatebe," as is perhaps a closer approximation of her name (Gibson, 2024, 107)) affords a critical reading of the trials, linking colonialism and the slave trade from Barbados, Boston, to Salem. We begin, then, with the stereotypical "Hollywood" witch.

Making and Marketing Witches

The Salem Witch Museum sits on North Washington Square. It is a brick and brownstone Gothic Revival building, constructed in 1844–1846 as Salem's East Church (Salem: Still Making History, 2024). The building is blazoned in red light at night, offering an eerie visual display for passersby. We wrote of our experience touring the museum.

> We walk into a large, dark room, with seats surrounding a central circle with the names of those accused printed on it. The room lights turn off, with the circle emitting a red glow, listing names of those executed. The presentation begins, with lights focusing on a demon in the corner with glowing red eyes. A speaker narrates several stages of the trial. Mannequins light up around the room, portraying the likes of Giles Corey being pressed to death and Reverend George Burroughs before his hanging.
>
> After the demonstration, a guide breaks the visitors into two groups. We start in the gift shop, which includes descendant packets, biographical and court information for those accused, and standard "Hollywood" witch merchandise, like a *Hocus Pocus* tote bag featuring the three Sanderson Sisters. The bag reads, "There's a little witch in all of us."
>
> We move to the exhibit, *Witches: Evolving Perceptions*. The guide shows us negative depictions of the witch, including the evil witch in the *Wizard of Oz*. She then directs us to a stage with two Pagans, a white man and woman, offering a more historical understanding of magic use. Then it is on to the Witch Hunt Wall Project, which warns of present-day witch hunting, including during the red scare, Japanese internment, and the HIV/AIDS crisis. It offers a general formula for witch hunts: "fear + a trigger = a scapegoat."

The Salem Witch Museum does much to dispel the image of the stereotypical Hollywood witch, that of a woman flying on a broomstick and casting spells. It offers information on Paganism and warns of witch hunting as a general phenomenon. Similarly, the guide assured us that during the trials there were "no special effects, fireworks, or toads." Despite such explanations, however, the industry remains highly reliant on the marketing of the stereotypical broom-stick flying witch, creating a disjuncture between goals to commemorate and educate versus those to commercialize and entertain (DeRosa, 2009).[4] In this section, we explore the marketing of three witches: that of the broomstick-flying white witch, the Wiccan witch, and the demonic Black Voodoo witch. These three marketed witches, we find, tends to associate those accused with actual witchcraft, obscuring their innocence, while profiting off the allure of the supernatural.

The broomstick-flying witch appears all over the city. The official Salem police department badge is the silhouette of a woman with a witch's hat who is riding on a broomstick (Figure 1.1). *Salem News* has the same depiction as its logo: a silhouette of a woman wearing a pointy hat and riding a broomstick. Salem's Witch City Taxi includes on its vehicles a print of the body of a broomstick-flying witch, with backseat passengers meant to humorously fill in the top part of the picture. Witch City Plumbing and

FIGURE 1.1 Salem Police Cruiser

Heating adopts the same logo: pointy hat, broomstick-flying witch. The image is sold on merchandise, such as a magnetic salt and pepper shaker of the Green "Bad Witch" and the White "Good Witch." Even the aforementioned Salem Witch Museum adopts the pointy-hat, broomstick-holding witch as its logo (DeRosa, 2009, 172). It was certainly jarring when, after we watched the performance of the trials, we were presented in the gift shop with a handbag bearing the "Sanderson Sisters" of *Hocus Pocus* fame.

Hocus Pocus is a mainstay of the industry, both in merchandise and on ghost tours. On numerous outings, we visited the Salem City Hall where, as guides pointed out, Bette Midler sang "I Put a Spell on You" when portraying villainous Winifred Sanderson. On several walking tours we visited the Ropes Mansion and Garden. Originally owned by Nathanial Ropes, a guide explained, the mansion was mobbed in 1774 as Robes maintained allegiance to the English Royal Crown. Every night, our guide continued, there is a light left on by the spirit of daughter Abigail "Nabby" Ropes, who was burned to death in the house. Yet, it is also the spot, he noted proudly, of the Halloween party scene in *Hocus Pocus*. He added, "On Halloween, they decorate the whole thing to look like Allison's house from the movie." Another use of the popular media witch occurs when guides take visitors to the bronze statue of a broomstick-riding Elizabeth Montgomery, of *Bewitched* (1964–1972) series fame. The statue was donated in 2005 by TV Land in honor of the show's Salem Saga (figure 1.2). Our guide connected the show's witch, Samantha, to those accused when informing us that the statue, located at the corner of Washington and Essex

FIGURE 1.2 Elizabeth Montgomery Statue

Streets in downtown Salem, is stationed in front of a trial judge's home. There is "a witch on his lawn for all eternity," he commented, despite that as Puritans, the accused certainly would not have practiced such witchcraft.

While the accused did not practice witchcraft, our guides still spoke of their magical capabilities—further imagining them as wielding supernatural power. After explaining that there were "no special effects, fireworks, or toads," the same guide linked the 1648 killing of Margarat Jones in Connecticut to a subsequent tempest. She warned, "Mess with one witch, there's a problem. Mess with two, there's hell to pay." Similarly, in *Haunted Salem* (2011, 20), Rosemary Guiley dispels the notion of the accused as witches but implies their supernatural powers through cursing vengeance. When executed on September 22, 1692, the accused Wilmott Redd proclaimed, "This town shall burn!" She now appears in the Marblehead old cemetery, Guiley informs, and "[a]t night her angry ghost wanders the tombstones and streets nearby, shrieking her final curse, 'This town shall burn!'" (135). Guiley connects Redd's cursing to an actual fire, stating that "Salem did burn" when referring to the Great Salem Fire of 1914 which destroyed 1,376 buildings (135). She concludes that, while the residents "killed innocent people to avoid evil. . . . Ironically, Salem suffered anyway, through the curses of the innocent who were executed" (20). While disconnecting the innocent from the witch, the idea that the accused conjured tempests and produced fires does much to parallel supernatural imagery of witches at once presented by witch-hunters (Gibson, 2024, 56).

The second way witches are marketed in Salem is through the image of the Wiccan witch. With origins in the 1940s, Wicca is "a revival, or reemergence, of an ancient nature-religion," with practitioners utilizing magical circles, alters, candles, and chanting (Luhrmann, 1989, 45).[5] While Wicca can be a source of feminist empowerment (Gibson, 2024), the marketing of the religion in the tourism industry largely focuses on individual consumer self-betterment, as tourist shops sell spells for money, love, true justice, luck, and protection. As these magical beliefs no longer represent a social threat (Federici, 2004, 143), they thus bolster a capitalist industry bent on selling spiritual commodities (Carrette and King, 2004). As Carmen Maria Machado (2022, xi) has put it of the "girl boss" figure of the witch, "[C]apitalism has gotten a hold of her." To add, any association between Wicca and those accused, as Robert Gagnon (2021, xviii) warns, "is to believe—and give legitimacy to—the falsehoods and fantasies of the accusers." Even empowering merchandise links the fantastical broomstick-flying witch with those accused, including one T-shirt which pairs the broomstick-flying witch with text that reads, "Women Who Behave Rarely Make History."

The third type of marketed witch is that of the Black Voodoo practitioner. This is particularly through misrepresentations of enslaved Arawak woman Tituba.[6] Perhaps the most popular representation of Tituba as a Voodoo practitioner is in Arthur Miller's *The Crucible* (1953), which was regularly mentioned by guides. In the play, Miller describes Tituba as a "negro slave" who has "been chanting over a boiling kettle containing, among other things, a live frog, while the girls of Salem Village dance, one of them naked, in the dark forest" (Hansen, 1974, 10).[7] The children's book *The Salem Witch Trials (Blast Back!)* similarly links the girls' interest in fortune telling to Tituba practicing Voodoo (Ohlin and Simó, 2017). Such depictions ultimately erase Tituba's Arawak identity and reinforce Black stereotypes of Voodoo—as wild, sexual, and corrupting white youth (Kendi, 2016).[8]

In our tours, guides also spoke of Tituba as practicing Voodoo. One clarified that Tituba was Arawak but, as she absorbed African culture from those enslaved, she practiced Voodoo under the nose of Parris. He stated, "She would tell the children folklore, like fortune telling by reading egg whites," despite that such fortune telling was English white magic (Gagnon, 2021, 85). A plaque in the Witch Dungeon Museum, another common tourist site, also informs visitors that Tituba "loved to talk about the devil and his powers."

> A servant woman Tituba, with her husband John Indian, loved to talk about the devil and his powers, as they sat around the fire at the home of Reverend Samuel Parris. Were the 9 girls, ages 9 to 19, really bewitched by Tituba's tales or did Tituba feed them hard cider or narcotics brought to Salem from the West Indies? Possibly the girls went into tantrums and fits

because they were bored and wanted attention, or maybe the Devil really did come to Salem.

Further cashing in on the dark allure of Voodoo, Salem shops sell Voodoo dolls, a pattern we will discuss further in the following chapter.

Whether flying on a broomstick, practicing Wiccan spells, or enacting Black Voodoo rituals, the marketable image of the witch is alive and well in Salem. As a guide put it to us, "Hollywood loves a Salem witch, yes they do." We would argue that Salem also loves a Hollywood witch, revealing much more of a feedback loop than a unidirectional influence (Ferrell et al., 2015). This sensationalist marketing, it bears repeating, is in tension with contrasting efforts to demystify notions of the accused as having anything to do with witchcraft. Not just the witch, but Salem also markets punishment in a way that is easily digestible—and, thus, void of ways of thinking about state violence as it occurred and is enacted today.

Making and Marketing Punishment

The Witch Dungeon Museum is another key tourist destination in Salem. It was opened to the public in 1972 and is not the actual dungeon but a re-creation with proper dimensions. We wrote of our experience,

> We walk inside and are welcomed to pews in front of a stage. Outlining the room are plaques that describe the witch trials in thirteen parts. Mannequins of judges and the jury fill the stage, outfitted in Puritan black-and-white dress. We watch a live reenactment of Rebecca Nurse being accused by Ann Putnam. The scene, the host tells us, uses actual dialogue from the trial. At one point, Putnam cries for help as she is putatively being attacked by spectral birds sent by Nurse.
>
> We move to the next part of the tour, the dungeon. About thirty of us fill a room surrounded by cells. The guide sets the scene, informing us that those incarcerated would be shackled, sometimes two hundred at a time. They even had to pay for their shackles. The dungeon regularly flooded with water reaching their waists; rats were prominent, and disease was rampant. We imagine being submerged under water in what already feels like a crowded space.
>
> There is a wooden beam from the original dungeon sitting atop a shelf. At intermission, our guide lets us look around and encourages we touch the beam. We place our hands on it, not really feeling much, but still imagining the horrors of being locked in the dungeon. We peak into cells, which include more mannequins, now languishing behind bars.
>
> Once finished, we go into a hallway. There are further cells that display aspects of the trials, including Tituba interacting with several children. Before we move to the final room, our guide prefaces that we will see a

display of victims hung on Procter's Ledge. She recommends we look away if need be. We witness the hung mannequins as we file to the exit and back onto the city sidewalk.

As with the dispelling of stereotypical images of the witch, our guide engaged sensitively with the campaign of terror that was the witch hunts (Federici, 2018, 32).[9] One guide, who was a descendent of the trials, reminded our group that it was OK to wear a stereotypical pointy witch hat as long as we understood the weight of the history. "It is important to talk a little more about the sadness of victims," she told us. Standing in front of the Witch Dungeon Museum, a guide also affirmed the misogynist dimensions of the trials, citing that those accused were often women, over 40 years of age, did not have children (or sons at least), and did not have a husband or brother to vouch for them.[10]

Yet, like the image of the witch, punishment in Salem is marketed within the profit-seeking tourism industry. It is done so in two key ways, as we discuss in this section. First is the marketing of victims as ghosts. Primarily, while guides inform on the lives of the accused and circumstances of their deaths, they also commodify death by telling of victims' ghosts haunting commercial establishments. Second is the marketing of state violence. Violence perpetrated by state actors, such as judges and local police, is engaged in the industry in a way that feeds audience spectatorship (Brown, 2009). The trials are also deemed a result of a few bad actors or collective hysteria, and the legal mechanisms that constituted such violence are largely disconnected from state violence today.

To the first point, the ghosts of those executed prove highly profitable in the Salem tourist industry. Consider Bridget Bishop, the first victim to be hanged in Salem on June 10. Bishop was tried on June 2 after being accused as a witch by numerous residents (Roach, 2013, 21–26). Six days later, Chief Justice William Stoughton signed her death warrant, ordering her to be "hanged by the neck until she be dead" (Rosenthal, 2009, Doc. 313). Part of her being targeted may have been due to her owning a business, being married three times, and given her manner of dress. While she "was not necessarily flashy," Daniel Gagnon (2021, 169) points out, "she was not a shy character either." Regarding the popular opinion against Bishop, Reverend Cotton Mather (1693, 223) had proclaimed upon her death, "There was little occasion to prove the witchcraft, it being evident and notorious to all beholders" (Gagnon, 2021, 169–170).

Several guides introduced Bishop when in an alley on 43 Church Street, the location where she owned an apple orchard. Guides relayed that she was targeted due to her not fitting the norms of Puritan womanhood. She wore a yellow dress that hung above her ankles, one explained. Yellow, he added, was associated with the devil. Bishop, guides pointed out, also operated a tavern and apple orchard, and she had been married three times. In this regard, they framed Bishop as a figure of women's empowerment. Yet, guides also made sure to flavor their stories by adding encounters people had with Bishop's ghost. After offering Bishop's story, one guide pointed to neighboring Turner's Seafood restaurant, which also sits on the property she owned. He told the story of a young employee who saw a ghost, screamed, ran away, and did not go back for her paycheck. It was presumed to be Bishop's ghost. Such stories operate, both, to add a thrilling dimension to the tours as well as to encourage patronage in local establishments.

Bishop's ghost also appears in promotional material and in popular media, encouraging prospective visitors to travel to Salem to catch a glimpse of her. Blogger devadmin writes for Salem Ghost Tours (2024) that in Turner's Seafood "several visitors have reported witnessing a woman in a long, white gown drifting throughout," and the aroma of apples has been noticed in the building. As told by Ghost City Tours, a former employee, Terri Colbert, shared for the History Channel's *Haunted History* that she saw a woman in a 17th-century white dress that may have been Bishop:

> It was a busy night. When I came up the stairs and looked up, I saw another woman standing on the other staircase leading up to the loft. I was petrified. My initial thought was that it was a person breaking into the restaurant. When I realized she wasn't a regular person, I ran back downstairs and almost fainted
>
> *(see also Baltrusis, 2014, 89).*

A crew member for Travel Channel's *Ghost Adventures* (2011) similarly felt when exploring the location that they "picked up what they thought was proof that they were interacting with the ghost of a woman put to death on charges of witchcraft." Like Turner's Seafood, Hawthorne Hotel also sits atop her phantom orchard. As Sam Baltrusis (2014, 89) explains in *Ghosts of Salem*,[11] some have smelled apples and witnessed "an apparition of a woman roaming the halls."[12]

Not just her ghost, but Bishop's trial and subsequent execution, is marketed. In *The Making of Salem*, Robin DeRosa (2009, 153–154) participated in *Cry Innocent*, a live reenactment of Bishop's witchcraft examination. The brochure advertises audience participation, "It's April 1692. Bridget Bishop is on the witness stand, and *YOU* are on the jury. Play your part in history." It adds of the autonomy afforded to the audience,

> As a member of the jury, you may cross-examine witnesses, argue with the defendant or give testimony yourself. Our actors will respond to your comments in character, revealing much about the Puritan mind.

DeRosa's group determined that, indeed, Bishop was guilty, with one man shouting, "Hang Her!" (154). Such interactive theater therefore allows consumers to ponder, play with, and even peer into their own, or cultural, violent sensibilities through the guise of entertainment and education. Indeed, the Witch Dungeon Museum gift shop sells certificates for $2 which, in jest, allow purchasers to "Accuse Your Friends" of witchcraft. The listed person, the document informs,

FIGURE 1.3 "Accuse Your Friends" at the Witch Dungeon Museum

"has been accused of witchcraft and is hereby summoned to appear at Salem's Witch Dungeon for questioning" (Figure 1.3).

Another marketed victim of the trials is Giles Corey. As our guides recounted, on September 19, Sheriff George Corwin pressed Corey to death when placing rocks on him in attempt to gain a confession. After three days, Corwin died, with his last words cheekily being, "More weight." In their tellings, guides viewed Corwin as a particularly bad actor. As one stated, "If there is a villain, one of them is Corwin, as he was a nasty, vile, fellow who touched all people negatively impacted by the witchcraft." Through stories of visitors being scratched and choked by Corwin, punishment is also viewed as a thrilling experience, allowing tourists to not just judge the acts of Corwin but *feel* state violence in spectral form. For instance, one guide stated that a tourist who visited the George Corwin House received scratches on their back. Another told of visitors being choked at the Birch Hotel. We wrote,

> The guide tells us that Corwin was corrupt. He abused his power and would choke people in his basement until they were close to death and had to grasp for air. He was feared and even referred to as the "strangler." He

accused, arrested, and hung many. It is said, the guide adds, that "Corwin is buried on the grounds and still terrorizes today." When visiting the hotel, he continues, "visitors have felt fingers around their neck and have received scratches on their bodies."

In discussing Salem's campaign of terror as the result of single bad actors like Corwin, or as stemming from "collective hysteria" (Meisler, 2024) or abstract intolerance and fear (DeRosa, 2009, 172), the sociohistorical mechanisms of state violence—and how it persists today—are left unattended to. Indeed, far from the result of superstition, the trials operated within scientific and legal advances, and judges employed numerous judicial standards that exist today.[13] Bernard Rosenthal (2009, Doc. 22) states of the 1692 trial documents,

> Although popular images are those of a society in the grips of "hysteria," there is nothing in the judicial attention to order and detail to suggest the legal authorities behaved that way. Certainly there were disruptions in the court by the "afflicted," but these disturbances did not change the orderly bureaucratic handling of the cases
>
> *(see Gagnon, 2021, 159).*

Certainly, courts have since eliminated spectral evidence. Yet, the pains of long-term pretrial incarceration (Price, 2015), debts incurred from imprisonment (Sobol, 2015), sexually violent strip searches (Davis, 1992), and false confessions (Miller, 2021) resonate all too well today. Nevertheless, within the tourism industry, contemporary state violence is obscured, with Salem police even adopting the broomstick-flying witch on their badge. In turn, consumers participate in state violence as jurors or feel phantom violence as a thrilling, marketable experience (Brown, 2009).

When marketing punishment, like when marketing witches, there is a disjuncture. Guides offer the scope of punishment and information on the accused and discuss the misogynist dimensions of the hunts. Yet, they also market the ghosts of the accused in a way that drives up local businesses, with punishment being framed as a thrilling and marketable experience. And, by focusing on the witch trials in general, guides do not discuss America's horror stories of colonialism and slavery, save for one who stated when we visited the house of sea captain and slave trader Joseph White, that he made a fortune off "a terrible thing called slavery."[14] With a focus on witch hunts, and some obfuscation of slavery, the twin horrors of colonialism and slavery are left beyond Kris Manjapra's (2022, 4) ghost line. To this, we now revisit the story of Tituba, or "Tatebe" (see Gibson, 2024, 107).

Tatebe Awakening

In fiction novel *I, Tituba*, Maryse Condé (1992) writes "Tituba" as a Black woman from West Africa. In the novel, Tituba was born from an African woman, Abena, who was raped by an English sailor. Her mother was hanged after defending herself from further sexual violence. Tituba escapes the plantation and develops knowledge of magic and herbs from healer Mama Yaya, and, upon her death, she continues to seek her spiritual guidance. Tituba returns to slavery upon marrying John Indian, and their owner, Susanna Endicott, sells them to Samuel Parris. Parris takes the two to Boston, where Tituba is eventually accused of witchcraft, is incarcerated, accuses Sarah Good and Sarah Osborne, and is sold to a Jewish merchant, Benjamin Cohen d'Azevedo, who frees her and pays her way back to Barbados. Here, she plots a revolt but is caught and executed by white planters, an execution that echoes her mother's death. Yet, her spirit lives on, "behind every revolt. Every insurrection. Every act of disobedience" (175). While not the Tituba of historical record, Condé presents her through multiple forms of Black struggle and resistance. She offers an empowering depiction, challenging notions of Tituba as demonic or simple-minded. Condé explains of her approach,

> For me *Tituba* is not a historical novel . . . I wanted to turn Tituba into a sort of female hero, and epic heroine, like the legendary, "Nanny of the maroons." . . . The result is a sort of mock-epic character
> *(Scarboro, 1992, 200–201; Jalalzai, 2009, 413).*

Angela Davis (1992, xi–xii) praises the author's approach, writing, "Tituba leaps into history, shattering all of the racist and misogynist misconceptions that have defined the place of black women" (Jalalzai, 2009, 421).

While certainly Condé's Tituba is not a demonic Voodoo figure, her characterization is not without criticism. Jane Moss (1999, 16) argues that "by knowingly or unknowingly transforming Tituba from an Indian to an African slave," Condé "substitutes one erasure for another." While we do see value in such fiction writing, as it portrays historical and symbolic truths of Black struggle (Johnson, 2001), the Tituba, or Tatebe, of historical record also speaks to multiple aspects of slavery, colonialism, racist ideas, and resistance, largely left out of Salem's ghost tourism industry. In this final section, we peer through the cracks of Salem's popular ghost stories, conjuring through Tatebe the horrors of colonialism and slavery otherwise silenced. This exploration begins, not in Salem, but in South America.

Tatebe was born in South America, most likely in present-day Venezuela. She lived on communal land as part of the Arawak tribe. She was kidnapped as a child by European slave traders and brought to Barbados (Gibson, 2024). Settled in 1620, Barbados was one of the earliest English colonies. Barbados officials announced in 1636 that "*Negroes* and *Indians* that come here to be sold, should serve for Life, unless a Contract was before made to the contrary"

(Kendi, 2016, 43). Indeed, slave labor proved essential for the colony's growth, particularly in 1640, as sugar cultivation replaced tobacco, cotton, ginger, and foodstuff production (Breslaw, 1995, 37–38). Africans were primarily utilized for fieldwork and Indigenous people for household labor. This division of slave labor comports with "the myth of the physically strong, beastly African, and the myth of the physically weak Native American who easily died from the strain of hard labor" (27). Representing mutual dehumanization, Africans were considered "a brutish sort of people" (55), and clerks listed Indigenous people, who never exceeded 0.2 percent of the population,[15] as "it," "which," or "that" (Breslaw, 1995, 72). By the early 1650s, Barbados "was described as the richest spot in the New Worlds" (Beckles, 2001, 28).

Tatebe worked on a three-hundred-acre sugar farm and refinery owned by Samuel Thompson (Gibson, 2024, 112). She was one of 67 enslaved people and was most likely between nine and 14 years old, as she was listed as a "pick-annie" (Breslaw, 1995, 55). She may have also had a child named Violet (87). When exchanging assets in 1676, Thomson sold Tatebe to Cornish slave trafficker Nicholas Prideaux. While she was more than likely exchanged a few times, it is possible that Prideaux directly sold her to Parris, of Salem witch trials fame (Gibson, 2024, 113). Parris had moved from New England to Barbados in the mid-1670s to settle his family estate after his father's death. He inherited two plantations and 69 enslaved people (Breslaw, 1995, 59). He resided in Bridgetown as a credit agent for other sugar planters (22), and Tatebe would have been used for domestic labor (56). Given his failing business ventures and a predilection for New England, Parris moved to Boston in 1680, where he brought Tatebe and enslaved John Indian. They were, as Gibson (2024, 113–114) describes, "hustled onto a ship and endured a stormy, miserable voyage over gray seas, traveling 3,500 miles north."

In Massachusetts, slave labor proved essential, particularly for tobacco production. Sales rose from 10,000 pounds in 1619 to 38 million in 1700. The first documented enslaved person imported was an American Indian in 1637 after the Pequot War, and the second was an African man a year later. In 1641, Massachusetts became the first English colony to legalize slavery in North America. Boston was an international market, where ships carried goods and passengers, servants, and enslaved people from the West Indies, Southern Europe, and the British Isles (Breslaw, 1995, 65). Captives were taxed with horses and hogs (67), and, standing in the Bridgetown market area "was a large barred cage, fourteen feet long, used to house runaway servants and slaves" (36).[16] While only 3 percent of the Boston population was African, residents nevertheless relied on enslaved domestic labor (68). Within the Parris household, Tatebe was certainly put to work.[17] In her testimony, she "describes 'washing' rooms: mopping mud off floorboards; swabbing the elegant furniture to remove the ash breathed out by open fires" (Gibson, 2024, 116). She was also made to adopt the Puritan faith and was subjected to

"English styles and standards of decoration, food, drink, and clothing" (Breslaw, 1995, 58).

Puritan racist ideas justified such treatment of Black and Indigenous people. Puritans described the devil as Black, with one Puritan accuser describing the devil as "a little black bearded man" (Kendi, 2016, 61). During the trial of Rebecca Nurse, accuser Ann Putman claimed witnessing a "Black man" whispering in Nurse's ear (Rosenthal, 2009, Doc. 28; Gagnon, 2021, 111), and Abigail Williams and Mary Walcott spoke of having seen in Nathanial Ingersoll's tavern the phantoms of "a great black woman . . . and an Indian" (Rosenthal, 2009, Doc. 85; Gagnon, 2021, 142). Africans were also described in the *Boston News-Letter* as being "much addicted to Stealing, Lying and Purloining" (Kendi, 2016, 69), and Reverend Cotton Mather described Indigenous people as idle, practicing sorcery, and having "a notorious want of Family Discipline" (72). Mather furthermore spoke of the inferiority of Africans when advocating for the religious instruction of those enslaved in a 1689 sermon,

> Puritans must religiously instruct all slaves and children, the inferiors. But masters were not doing their job of looking after African souls, which are as white and good as those of other Nations, but are Destroyed for lack of Knowledge
>
> (59).

While an assimilationist approach ran against Barbados's and Southern planters' segregationist approaches, it nevertheless justified slavery and reinforced the idea of white superiority and Black inferiority. As they sought out witchcraft, Puritans perpetuated *race*craft when delineating racial superiority and inferiority (Fields & Fields, 2014).[18]

Yet, Tatebe's testimony also reveals resistance—that she was not a "simple slave" or "passive victim" (Breslaw, 1997, 549). While Tatebe was not native to the New England area, Indigenous people, it should be said, fought settlers wherever they were, including during King Philip's War (1675–1676; Cray, 2009). Native uprisings created in white settlers "an intense hatred and suspicion of all Indians" (Breslaw, 1995, 68). While Tatebe's confession was the catalyst for further accusations, she also demonstrated creative resistance. Combining English, African, and Indigenous forms of storytelling (xxii), including effigies representing guardian spirits as well as evil spirits, such as kenaimas who were real people "compelled to kill, cause sickness, or bring about other grave misfortune." (17–19). In her descriptions of a tall white man and a woman with a white silk hood in Boston, she directed examiners to "look among the elite for evil beings," perhaps even to Parris himself (119–120). As Elaine Breslaw (1997, xx) concludes of the complexity of Tatebe's testimony, "In the process of confessing to fantastic experiences, she created a new idiom of resistance against abusive treatment and inadvertently led the way for other innocents accused of the terrible crime of witchcraft."

Tatebe's fate is unclear. She spent over a year in the Salem jail before being incarcerated in Boston (Gibson, 2024, 124). She was not tried, with grand jury Abraham Haseltine writing "ignoramus," meaning "we don't know, we don't approve" (125). Tatebe might have been sold by Parris to pay off her jail fees, or she may have been "freed by the state or by well-wishers, of whom there were many by 1693" (127). A tour guide stated that, since there was no trace of her, perhaps she was the "only one able to find peace." While her fate remains uncertain, we can see her as an Arawak woman whose story touches on slavery, colonialism, racist ideas, and resistance not discussed in the Salem tourist industry.

Conclusion—Deluded by Whiteness

In May 1693, Governor William Phips pardoned those imprisoned on witchcraft charges. In October, the General Court introduced a bill for a day of fasting (Gagnon, 2021, 235). In 1697, Judge Samuel Sewall offered a formal apology, and 12 jurors came forward to beg for forgiveness, maintaining they were "deluded by Satan" (250). In 1702, the court declared the trials unlawful, and, in 1711, the colony restored the rights and good names of the accused and offered 600 pounds in restitution to their heirs (27). Memorials have been established to honor the innocent, including that for Rebecca Nurse in present-day Danvers in 1885 (179) and the Salem Witch Trial Memorial in 1992 located on Salem's Liberty Street. Salem Mayor Driscoll told *CBS Boston* in 2016 that the trials are "definitely a dark part of our history, an infamous time in Salem when people turned onto each other. I think we learned a lot of lessons and we've worked hard to overcome what happened in 1692."

Nevertheless, the legacy of the witch hunts continue—as an unresolved social violence (Gordon, 2011, 2) that is not reckoned with in the ghost tourism industry, even if it is touched upon. The misogyny of the witch hunts persists today in gender discrimination in political participation, employment, and medical care to name a few examples (Chollett, 2022). Witch hunts still occur, for instance, in Tanzania, Nepal, Mozambique, Ghana, Papua New Guinea, Saudi Arabia, and India (Federici, 2018, 4). Of note, demonology came not from an inherent African "backwardness" but was incorporated by Christian missionaries, who introduced the notion of the Christian devil and rendered Indigenous beliefs and spiritual practices as demonic (Gibson, 2024, 207–208). Contemporary witch hunts are furthermore a response to, as Federici (2018, 51) explains, debt bondage and the enclosures of communal land through the adoption of economic policies designed by the World Bank and International Monetary Fund (IMF). As Gibson (2024, xvii) writes on the unresolved violence of the hunts, "women of all kinds were overwhelmingly the victims of witch trials, and that misogyny still haunts global cultures."

The scapegoating of populations, even through religious zealotry, furthermore remains pervasive. Just consider the U.S. 1980s satanic panics over fantasy tabletop board games (Laycock, 2015) and daycare providers (de Young, 2003).

As Gibson (2024, xvii) reminds, scapegoating has also historically included Jewish and Muslim people, and, by the late 19th century, included "spiritualists, anarchists, communists, suffragists, or homosexual people; in the twentieth century, civil rights campaigners and anticolonial nationalists joined the list." Indeed, the U.S. wars on drugs, crime, and terror has associated marginalized groups with othered social categories, such as the "drug dealer," "criminal," and "terrorist." As such, witch hunting, Mark Neocleous (2016, 97) contends, "is a permanent feature of the reproduction of capital."

Salem continues to benefit off the witch trials through a mixture of education, sensationalism, and commercialization, all within a lucrative ghost tourism industry. In marketing witches and punishment, tour guides largely overlook the impact of slavery and colonialism in the area. Tatebe, however, also offers a way to retrieve and place these American horrors into historical memory: affording a new kind of "spectral evidence," not as evidence of the spectral powers of the witch, but as evidence of the specters of the past which haunt us today. In continuing our hauntologically informed examination of space (Fiddler, 2019, 474), we travel to New Orleans, Louisiana, which was the largest slave trade hub in the South.

Notes

1 Their relationship is uncertain, although "niece" was "a more generic term for a familial relationship" (Gagnon, 2021, 79).
2 Early writers pointed to familial divides (Boyer and Nissenbaum, 1974) and even delusion-causing ergot poisoning (Caporael, 1976).
3 As the great-great grandson of Judge John Hathorne, Nathaniel Hawthorne (1851) grabbles with his family heritage in *House of the Seven Gables*. The novel tells of the landgrab of Matthew Maule's property, and, as Colin Dickey (2016, 26) observes, "it is precisely because the house is the physical spoils of this injustice that it becomes haunted."
4 The marketing of the witch trials in Salem has a long history (DeRosa, 2009, 157). In 1891, a souvenir spoon with the "Witch City" design was sold by Daniel Low & Co. (Salem Witch Museum, 2024).
5 The term "Wicca" was introduced in 1954 by Gerald Gardner (2004) in *Witchcraft Today*. Salem is a popular destination for those who practice the religion, with one tour guide confirming that, while there was no witchcraft at the time, "twenty percent are here, and I am glad for that." As many as 5,000 residents of Salem identified as practicing witches in 2021 (Burton, 2017).
6 In 17th-century court documents, Tituba was listed as "Titibe an Indian Woman," "Titiba an Indian woman," and "Tituba Indian" (Hansen, 1974, 4).
7 In the play *Giles Corey*, Henry Wadsworth Longfellow (1868) also lists Tituba as "half-Indian, half-negro," with a "black and fierce" father. In *The Devil in Massachusetts*, Marion L. Starkey (1950, 9–11) describes her as lazy and "half savage," with "slurred southern speech and tricksy ways." As Starkey's continues, she prefers "idling with the little girls" to working, and it is such "idling" that "Tituba yielded to the temptation the children tricks and spells, fragments of something like voodoo remembered from the Barbados" (paraphrased by Hansen, 1974, 8).
8 To remind, the baking of the witch's cake, too, was not Voodoo but an English folk tradition. As Daniel Gagnon (2021, 85) points out, the cake "was the closest thing to

witchcraft done in Salem Village in 1692, and it was done by a church member under the minister's own roof."

9 Although the English Witchcraft Act of 1604 outlawed the use of torture to extract confessions (Gagnon, 2021), and thus was adopted in Salem, those accused nevertheless faced troubling conditions. Rebecca Nurse, for instance, sat for months in the Salem and Boston jails, where she was weighed down with iron shackles used to presumably prevent her from sending out specters (Gagnon, 2021, 138).

10 While age demographics were not cited in the tours, it is important to note that the majority of accused were middle-aged women (40 to 59), not older-aged women (60 and over; Gagnon, 2021, 106).

11 A guide claimed to DeRosa (2009, 168) that she experienced Bishop's ghost in the Witch Dungeon Museum, where she had been working for 16 years. One night, she heard a woman humming after hours. It was on the anniversary of Bishop's hanging, so the guide presumed the voice to be hers. This was despite that the Witch Dungeon Museum is a recreation of the actual dungeon, with the original dungeon sitting 500 feet away.

12 A variety of businesses profit from the haunted aesthetic stemming from the witch trials. This includes the Rockafellas Restaurant. Employees took a picture and saw "a figure of a lady in a blue dress." The owners created a mural for the "lady in blue" and now offer a specialized cocktail drink. ABC reports of the sighting and their profitability, "It's safe to say she's brought the two good fortune over the past couple of decades, while haunting patrons in the process" (Jagolinzer, 2023). Other establishments include Boris Karloff's Witch Mansion and Salem's Museum of Myths and Monsters: Terror on the Wharf (DeRosa, 2009, 175).

13 Isaac Newton, Francis Bacon, John Locke, and Thomas Hobbes all believed in witchcraft (Goode and Ben-Yehuda, 2010, 170).

14 A focus on abolitionist influence, without larger inquiry, can present Salem as otherwise racially progressive. Guides discussed, for instance, Fredrick Douglass speaking at the Lyceum Hall of the Salem Lyceum Society, which sits atop Bishop's apple orchard. To add, the Salem tourism industry also valorizes the city's history of fishing and maritime trade with Europe, the East Indies, and the West Indies (DeRosa, 2009, 156). Standing in front of the City Hall, our guides regularly boasted that, not only did Bette Midler sing there but it is where America's first millionaire lived, the merchant Elias Hasket Derby. Yet, such a focus on international trade downplays the role of slave labor in making tradeable goods, such as sugar, molasses, and coffee.

15 By 1679, the European population size was 21,725 and the African population size was 32,473 (Breslaw, 1995, 35).

16 Indeed, in 1674, a royal proclamation by the governor of Massachusetts Bay Colony, John Leverett, granted the Royal African Company exclusive rights in the slave trade (Love, 2022, 49).

17 While it was not permitted to formally enslave Indigenous people in Massachusetts, Tatebe was functionally treated as such (Gibson, 2024, 114).

18 To add, these racist ideas, like the demonology of the witch hunts, are not born from mere prejudice or ignorance but are deliberately constructed in law, religion, and scientific inquiry (Roberts, 2022). Indeed, humanitarian efforts had also been articulated for the abolition of slavery. Four-years before the Salem Witch Trials and a year before for Mather's 1689 sermon, the 1688 German Quacker Petition against Slavery in Pennsylvania was published. It was authored by four abolitionists, including Gerret Hendericks, Derick up de Graeff, Francis Daniell Pastorius, and Abraham op den Graeff, and it was, according to Katharine Gerbner (2018, 70), "one of the first documents to make a humanitarian argument against slavery." (See Lebron, 2022, 63).

References

Baltrusis, Sam. 2014. *Ghosts of Salem: Haunts of the Witch City*. Mount Pleasant: Arcadia Publishing.

Beckles, Hilary. 2001. "The 'Hub of Empire': The Caribbean and Britian in the Seventeenth Century." In *The Oxford History of British Empire: Volume 1: The Origins of Empire: British Overseas Enterprise to the Close of the Seventeenth Century*, edited by Nicholas Canny, 218–240. Oxford: Oxford University Press.

Ben-Yehuda, Nachman. 1980. "The European Witch Craze of the 14th to 17th Centuries: A Sociologist's Perspective." *American Journal of Sociology*, 86(1): 1–31.

Boyer, Paul, and Stephen Nussbaum. 1974. *Salem Possessed: The Social Origins of Witchcraft*. Cambridge: Harvard University Press.

Breslaw, Elaine G. 1995. *Tituba, Reluctant Witch of Salem: Devilish Indians and Puritan Fantasies*. New York: New York University Press.

Breslaw, Elaine G. 1997. "Tituba's Confession: The Multicultural Dimensions of the 1692 Salem Witch-Hunt." *Ethnohistory*, 44(3): 535–556.

Brown, Michelle. 2009. *The Culture of Punishment: Prison, Society, and Spectacle*. New York: New York University Press.

Burton, Tara Isabella. 2017, October 30. "There Weren't Any Witches in Salem in 1693. But There Sure Are Now." *Vox*. vox.com

Caporael, Linda R. 1976. "Ergotism: The Satan Loosed in Salem? Convulsive Ergotism May Have Been a Physiological Basis for the Salem Witchcraft Crisis in 1692." *Science*, 192(4234): 21–26.

Carrette, Jeremy, and Richard King. 2004. *Selling Spirituality: The Silent Takeover of Religion*. New York and London: Routledge.

CBS Boston. 2016, January 14. "Actual Site of Salem Witch Hangings Discovered." *CBS News*. https://www.cbsnews.com.

Chollet, Mona. 2022. *In Defense of Witches: The Legacy of the Witch Hunts and Why Women Are Still on Trial*. New York: St. Martin's Press.

Cray, Robert E. 2009. "'Weltering in Their Own Blood': Puritan Casualties in King Philip's War." *Historical Journal of Massachusetts*, 37(2):106–123.

Condé, Maryse. 1992. *I, Tituba, Black Witch of Salem*. Charlottesville: University of Virginia Press.

Davis, Angela Y. 1992. "Foreword." In *I, Tituba, Black Witch of Salem by Maryse Condé*, xi–x. Charlottesville: University of Virginia Press.

devadmin. 2024. "Turner's Seafood at Lyceum Hall." Salem Ghost Tours. https://salemghosts.com.

de Young, Mary. 2003. *The Day Care Ritual Abuse Moral Panic*. Jefferson: McFarland.

DeRosa, Robin. 2009. *The Making of Salem: The Witch Trials in History, Fiction and Tourism*. Jefferson: McFarland.

Federici, Silvia. 2004. *Caliban and the Witch*. New York: Autonomedia.

Federici, Silvia. 2018. *Witches, Witch-hunting, and Women*. Binghamton: PM Press.

Ferrell Jeff, Keith Hayward and Jock Young. 2015. *Cultural Criminology: An Invitation* (2nd ed). Los Angeles: Sage Publications.

Fiddler, Michael. 2019. "Ghosts of Other Stories: A Synthesis of Hauntology, Crime and Space." *Crime, Media, Culture*, 15(3): 463–477.

Fields, Karen E., and Barbara J. Fields. 2014. *Racecraft: The Soul of Inequality in American Life*. New York: Verso.

Gagnon, Daniel. 2021. *A Salem Witch: The Trial, Execution, and Exoneration of Rebecca Nurse*. Yardley: Westholme.

Gardner, Gerald Brosseau. 2004. *Witchcraft Today*. New York: Citadel Press.

Gerbner, Katharine. 2018. *Christian Slavery: Conversion and Race in the Protestant Atlantic World*. Philadelphia: University of Pennsylvania Press.

Ghost Adventures. 2011. Salem Witch House/Lyceum Restaurant.

Gibson, Marion. 2024. *Witchcraft: A History in Thirteen Trials*. New York: Simon & Schuster.

Goode, Erich, and Nachman Ben-Yehuda. 2010. *Moral Panics: The Social Construction of Deviance*. Hoboken: John Wiley & Sons.

Gordon, Avery. 2011. "Some Thoughts on Haunting and Futurity." *Borderlands*, 10(2): 1–21.

Guiley, Rosemary Ellen. 2011. *Haunted Salem: Strange Phenomena in the Witch City*. Mechanicsburg: Stackpole Books.

Hansen, Chadwick. 1974. "The Metamorphosis of Tituba, or Why American Intellectuals Can't Tell an Indian Witch from a Negro." *New England Quarterly*, 3–12.

Hawthorne, Nathaniel. 1851. *The House of the Seven Gables*. Washington: Charles E. Merrill Company.

History Channel 2. 2001. New England. *Haunted History*.

Hutton, Ronald. 2017. *The Witch: A History of Fear, From Ancient Times to the Present*. New Haven: Yale University Press.

Kendi, Ibram X. 2016. *Stamped from the Beginning: The Definitive History of Racist Ideas in America*. Boston: Bold Type Books.

Jagolinzer, Jordyn. 2023, October 31. "Rockefellas Restaurant a Popular Part of Salem's Spooky History." *CBS News*. https://www.cbsnews.com.

Jalalzai, Zubeda. 2009. "Historical Fiction and Maryse Condé's *I, Tituba, Black Witch of Salem*." *African American Review*, 43(2/3): 413–425.

Johnson, Walter. 2001*Soul by Soul: Life Inside the Antebellum Slave Market*. Cambridge and London: Harvard University Press.

Laycock, Joseph P. 2015. *Dangerous Games: What the Moral Panic over Role-Playing Games Says about Play, Religion, and Imagined Worlds*. Berkeley: University of California Press.

Lawson, Deodat. 2002. "A Brief and True Narrative of Witchcraft at Salem Village." In *Narratives of the New England Witchcraft Cases*, edited by George Lincoln Burr. Mineola: Dover Publications.

Lebron, Christopher J. "The Germantown Petition against Slavery." 2022. In *Four Hundred Souls: A Community History of African America, 1619–2019*, edited by Ibram X. Kendi and Keisha N. Blain, 62–64. New York: One World.

Longfellow, Henry W. 1868. *Giles Corey of the Salem Farms*.

Love, David A. 2022. "The Royal African Company." In *Four Hundred Souls: A Community History of African America, 1619–2019*, edited by Ibram X. Kendi and Keisha N. Blain, 47–50. New York: One World.

Luhrmann, Tanya M. 1989. *Persuasions of the Witch's Craft: Ritual Magic in Contemporary England*. Cambridge and London: Harvard University Press.

Machado, Carmen Maria. 2022. "Foreword." In *In Defense of Witches: The Legacy of the Witch Hunts and Why Women Are Still on Trial*, edited by Mona Chollet, v–viii. New York: St. Martin's Press.

Manjapra, Kris. 2022. *Black Ghost of Empire: The Long Death of Slavery and the Failure of Emancipation*. New York: Simon & Schuster.

Mather, Cotton. 1693. *The Wonders of the Invisible World*. London: John Russell Smith.

McCarthy, Niall. 2019. "The Death Toll of Europe's Witch Trials." *Statista*. https://www.statista.com.

Meisler, Alexa. 2024, April 2. "13 Most Haunted Places in Salem." *52 Perfect Days*. https://52perfectdays.com.

Miller, Arthur. 1953. *The Crucible*.

Miller, Reuben Jonathan. 2021. *Halfway Home: Race, Punishment, and the Afterlife of Mass Incarceration*. New York: Little, Brown and Company.

Moss, Jane. 1999. "Postmodernizing the Salem Witchcraze: Maryse Condé's *I, Tituba, Black Witch of Salem*." *Colby Quarterly*, 35(1): 3.

Neocleous, Mark. 2016. *The Universal Adversary: Security, Capital and "the Enemies of All Mankind."* New York and London: Routledge.

Ohlin, Nancy, and Simó, Roger. 2017. *The Salem Witch Trials (Blast Back!)*. New York: Little Bee Books.

O'Reilly, Jennifer. 2019. "'We're More Than Just Pins and Dolls and Seeing the Future in Chicken Parts': Race, Magic and Religion in *American Horror Story: Coven*." *European Journal of American Culture*, 38: 29–41.

Peabody, Sue. 2002. *"There Are No Slaves in France": The Political Culture of Race and Slavery in the Ancien Régime*. Oxford: Oxford University Press.

Price, Joshua M. 2015. *Prison and Social Death*. New Brunswick, NJ: Rutgers University Press.

Roach, Marilynne. 2004. *The Salem Witch Trials: A Day-by-Day Chronicle of a Community Under Siege*. Lanham, MD: Taylor Trade Publishing.

Roach, Marilynne. 2013. *Six Women of Salem: The Untold Story of the Accused and Their Accusers in the Salem Witch Trials*. Cambridge: Da Capo Press.

Roberts, Dorothy E. 2022. "Race and the Enlightenment." In *Four Hundred Souls: A Community History of African America, 1619–2019*, edited by Ibram X. Kendi and Keisha N. Blain, 119–122. New York: One World.

Rosenthal Berndard. 2009. "General Introduction." In *Records of the Salem Witch-Hunt*. Cambridge: Cambridge University Press.

Salem: Still Making History. 2024, January 29. "50 Years of the Salem Witch Museum." *Destination Salem*. https://www.salem.org.

Salem Witch Museum. 2024. "Daniel Low Reproduction Spoon." https://salemwitchmuseum.com.

Scarboro, Ann Armstrong. 1992. "Afterword." In *I, Tituba, Black Witch of Salem*, by Maryse Condé. Trans. Richard Philcox. Charlottesville, VA: University Press of Virginia.

Sobol, Neil L. 2015. "Charging the Poor: Criminal Justice Debt and Modern-Day Debtors' Prisons." *Maryland Law Review*, 75.2: 486–540.

Starkey, Marion L. 1950. *The Devil in Massachusetts: A Modern Inquiry into the Salem Witch Trials*. Potomac: Pickle Partners Publishing.

2

SOUL SEARCHING IN NEW ORLEANS

Introduction

"In short, living in New Orleans in the 1800s, in days before air conditioning and modern sanitation, on the whole was pretty miserable," as Kathryn Olivarius (2022) describes to Jonathan Van Ness on the *Getting Curious* (2023) podcast. New Orleans was founded in 1718 by French Jean-Baptiste Le Moyne de Bienville on land originally occupied by the Chitimacha Indians. The city sits between the Mississippi River and Lake Pontchartrain, making it an ideal trading post. Yet, to Olivarius's point, conditions were less than ideal. The city was in the middle of a swamp and was subject to periodic flooding from the Mississippi River (Long, 2007, 4). It was "plagued by insects, tropical diseases, poor sanitation, and food shortages" (4). Over 41,000 people died from yellow fever between 1817 and 1905 (Louisiana Office of Public Health, 1934). Residents were subjected to hurricanes and thunderstorms from the Gulf of Mexico, and, in 1788 and 1794, two fires destroyed nearly two-thirds of the city, as the structures were made of wood before being rebuilt by the Spanish. More recently, in 2005, Hurricane Katrina flooded 80 percent of the city. As one resident said on the 10-year anniversary of Hurricane Katrina, "Even cities feel trauma"—and New Orleans is no exception (Buncombe, 2015; Dickey, 2016, 247).

Given New Orleans's history of death and disaster, ghost stories abound. In *Southern Ghost Stories*, Allen Sircy (2023, 13) states of the two fires, "The scars left by the fires seem to have etched themselves into the fabric of the neighborhood, inspiring tales of restless spirits and eerie occurrences." In *New Orleans Ghosts, Voodoo and Vampires*, Kalila Smith (2016, 4) elaborates on how the city's history of "violent death" and "strong emotion" also contributes to ghostly activity:

DOI: 10.4324/9781003397809-3

Murderers, thieves, rapists and common criminals were among the first to populate the area. Living conditions were deplorable. Harsh elements, quicksand, alligators, venomous snakes, mosquitoes and disease were rampant. The murder rate was high. Add a couple of major fires that devoured the city and many of its inhabitants, numerous hurricanes, wars, and more than twenty-seven yellow fever epidemics over the next two centuries and you have excellent conditions for ghostly activity.

To experience this ghostly activity firsthand, companies offer candlelight tours of cemeteries, walking tours of the French Quarter, entry into haunted buildings, and bus tours in and out of the city. "I don't care who you are, a ghost hunter, tour guide, or bartender," one guide told our group, "Everyone here has seen something. The whole French Quarter is haunted." Another put it succinctly, "New Orleans is the most haunted city because of one word: death."

Not just disease and disaster, but New Orleans also held the largest slave market in the South. The first Africans were imported into Louisiana in 1719, and slavery was maintained during French and Spanish rule. The trade flourished with the 1803 incorporation of Louisiana into the United States. Louisiana's first territorial governor, William C.C. Claiborne, reported to President Thomas Jefferson of the wealth garnered from slave labor in sugar production, writing that the "facility with which the sugar Planters amass wealth is almost incredible" (Follett, 2005, 18). One in three residents had been enslaved, and 135,000 people were purchased and sold in the city, with researchers identifying 52 distinct sites of the trade (Chavez, 2022). As the Historic New Orleans Collection outlines,

> Auction blocks in the sumptuous rotunda of the St. Louis Hotel, private residences, public parks, decks of ships moored along the Mississippi, high-walled slave pens, and commercial complexes such as Banks Arcade all served as sites for the buying and selling of human beings.

When enslaved people were sold from the Upper to the Lower South of the United States, often the first stop, as Eve Abrams (2015) details, "was the slave markets of New Orleans, where families were divided for good." Often unbeknownst to the average tourist swaying down Bourbon Street, the city's atmosphere is built on the backs of enslaved Black people. How is the history of slavery in New Orleans discussed in the city's ghost tourism industry?

In this chapter, we explore how slavery is represented in three popular ghost stories: of Madame Delphine LaLaurie, who conducted "Frankenstein"-like experiments on those she enslaved; of Julie, a woman of color who entered a relationship with a white Frenchman at a plaçage party; and of Marie Laveau, the Voodoo queen of New Orleans who is both demonized and valorized in the city. We find that these stories downplay the violence of slavery in New Orleans, while reinforcing anti-Black stereotypes: they portray brutalized enslaved masses, adopt the tragic mulatto trope, and demonize and commercialize Voodoo (see Miles, 2015; Foley, 2021). By contrast, we recommend seeking markers of slavery in the city that do not exploit Black suffering and reframe Voodoo as a powerful force for enslaved resistance. We begin at the Haunted House on Royal Street, where slave owner Madame Delphine LaLaurie resided.

LaLaurie and the Brutalized Enslaved

The LaLaurie mansion is a tourist hotspot. The building, as was described in the *Daily Picayune* (1892), towers "high above every street in the French Quarter . . . [it is] a large, solid rectangular mass, with its three stories and attic and gray stuccoed front and sides" (Taylor, 2010, 61; Figure 2.1). Here, Delphine LaLaurie hosted lavish parties, which she would at times excuse herself from to torture those she enslaved as parties went on below (Taylor, 2010, 68). Her abuse was most evident when a fire was set in her home on the morning of Thursday, April 10, 1834.

The fire, according to the *Courier* (1834) on April 10, began in the kitchen and was "soon wrapped in flames." Judge Jacques François Canonge gave orders to break down the doors of the slave quarters above the kitchen.

FIGURE 2.1 LaLaurie's Mansion

A group of citizens entered the service wing and "were greeted by an 'appalling sight,' as 'several wretched negroes' emerged from the smoky interior, 'their bodies covered with scars and loaded with chains.'" On April 11, the *Bee* (1834) accounted of the scene,

> Seven slaves, more or less horribly mutilated, were seen suspended by the neck, their limbs apparently stretched and torn from one extremity to the other. . . . The slaves were the property of the demon in the shape of a woman. . . . Language is powerless and inadequate to give proper conception to the horror which a scene like this must have inspired.

LaLaurie escaped her assailants in carriage, while a crowd "of all classes and colors" tore up her house (*Courier*, 1834, April 11), with a loss of property estimated at $10,000 (see Long, 2012, 90–95).[1]

As is evident by local reporting and eyewitness accounts, LaLaurie's treatment of those she enslaved was nothing short of horrific. French Consul to New Orleans Armand Saillard sent a report to the minister of foreign affairs citing "dislocated heads, the legs torn by the chains, and the bodies streaked [with blood] from head to foot from whiplashes and sharp instruments" (Long, 2012, 97). Yet, subsequent ghost stories of the event added further gory details, such as in Jeanne DeLavigne's (1946) *Ghost Stories of Old New Orleans* and the aforementioned *New Orleans Ghosts, Voodoo and Vampires* (Smith, 2016). In the latter, Smith (26–27) writes what would become the general description of the scene by tour guides,

> Slaves were chained to the walls throughout, maimed and disfigured, obviously victims of cruel medical experiments. . . . Many looked dead but some were still alive. Several had faces so disfigured they looked like gargoyles. One man looked as if he has been the victim of some crude sex operation. . . . Another victim obviously had her arms amputated and her skin peeled off in a circular pattern, making her look like a human caterpillar. Yet another had been locked in a cage that a newspaper described as barely large enough to accommodate a medium size dog. Breaking the cage open, the rescuers found that all of her joints had been broken and reset . . . at odd angles so she resembled a human crab.

We heard such accounts numerous times by guides, of crablike disfigurement, sex operations, and of LaLaurie drilling a hole in the head of one enslaved person. While such storytelling does demonstrate the brutality of slavery, and was adopted by abolitionist Harriet Martineau (1838), it also does much to sensationalize and commercialize Black suffering, while reinforcing numerous anti-Black stereotypes.

The imagery presented in the story, first and foremost, associates Blackness with a kind of beastliness. The notion of Africans as beasts, or animals more broadly, was foundational to proslavery rhetoric. As early as the 15th century, to justify the Portuguese slave trade, Prince Henry's biographer Zarura asserted that Africans "lived like beasts without any custom of reasonable beings" (Kendi, 2016, 24). Not just during tours, but such a description appears in *American Horror Story: Coven*. While hosting a party, LaLaurie, played by Kathy Bates, learns that one of her daughters is having sexual relations with an enslaved man. In the attic, LaLaurie chains him, tortures him, and places upon him a bull's head, turning him into a hybrid beast: the minotaur (Miles, 2015, 76; Gilmore, 2012, 8). She states to enslaved Bastian before placing the head, "You want to rut like a beast, then we're going to treat you like one" (*American Horror Story*, 2013). Not just an "animalistic mass," as Laura Foley (2021, 68) contends, but those enslaved in the story are also "faceless." That is, the ghost stories of the brutalized enslaved often do not offer identifying information, that it appears there were seven carried out, including four women, two men, and one unidentified by gender. One may have been elderly Françoise, referred to in inventory records as "the old one" and "the hunchback" (Long, 2012, 101).

Stories of the brutalized enslaved, too, commercializes Black suffering (Hartman, 2022, 26). As Tiya Miles (2015, 62) reminds us, the "punishment visited upon the bodies of slaves is essential to the story's horrific allure." Through storied repetition, those LaLaurie enslaved can never truly be free of their punishment, so to speak. They are "doomed to be tortured for eternity," kept in perpetual suffering for the sake of horrific allure (Foley, 2021, 68). In *Southern Ghost Stories*, Allen Sircy (2023, 108–109), for instance, speaks of enslaved Lisette, who was assigned to work in the LaLaurie's kitchen and is "said to wander the mansion, forever trapped in a realm between life and death." In *Haunted New Orleans*, Troy Taylor (2010) tells of numerous enslaved spirits who are heard screaming from the house. One guide, as Miles (2015, 62) writes of her tour, spoke of a white male resident who "saw a black man's ghost near the former slave quarters. The ghost held his mouth open in a silent scream, revealing a nub in the back of his mouth where his tongue should have been." On top of this, the ritualistic telling of such horror tends to desensitize audiences to Black suffering. During a self-guided audio tour, the recorded voice appeared to, as we wrote in our field notes, "speed through the scene, offering masses of otherwise horrifying information. Without much else to offer, the disembodied voice directs us to the next spot."

Stories of LaLaurie, too, present her as exceptional in her cruelty, thus minimizing the horrors of slavery in New Orleans more generally. She has been described as "filthy rich" (Cable, 1889), as "sadistic," "insidious," and "maniacal" (Sillery, 2001; see Miles, 2015, 58), as "insane" and "cold blooded" (Taylor, 2010), and as a "demon in the shape of a woman" (*Bee*, 1834). These descriptions present LaLaurie as unique to slaveholding society in New Orleans, which had relied on Code Noir. The Code Noir, introduced in 1724, was a set of slave regulations that restricted punishment, banned separation of families, and prohibited labor on Sundays. In *American Horror Story: Coven*, a tour guide distinguishes Code Noir from LaLaurie's abuse:

The Code Noir, a decree that defined the conditions of slavery, didn't exist on these grounds. It was replaced by the Madame's own code of terror. And the torture she inflicted on her slaves would spawn one hundred seventy nine years of hauntings

(American Horror Story, 2013).

Despite this representation, the code did permit a variety of punishments, such as whipping and the cutting off of ears, and the prohibition against torture and murder often went unenforced (Miles, 2015, 74). The code also regulated for enslaved people weapon possession; congregation; selling items like sugar, fruits, vegetables, firewood, or herbs; and to travel without a written note, and it permitted death for an enslaved person striking a master, mistress, or child (Ralph, 2022, 59–60). To add, while LaLaurie may appear outside the norm of white Southern womanhood, Southern women routinely slapped, hit, and brutally whipped those they enslaved (Miles, 2015, 71–74). On top of this, when describing LaLaurie as exceptionally monstrous (King, 2017), her physician husband, Louis LaLaurie, is left unscrutinized, despite him having the medical experience to conduct such experiments (Love & Shannon, 2011, 35). Ultimately, in focusing on the monstrosity of LaLaurie, New Orleans is viewed as otherwise having been "a good place to have been a slave" (Miles, 2015, 71).

As a final note, our tour guides also presented key civil rights moments in the city, such as Homer Plessy challenging segregation on the East Louisiana Railroad in 1892 and American civil rights activist Ruby Bridges paving the way for school desegregation in 1960. While these certainly should be mentioned, without scrutiny of extant systemic racism, they may present New Orleans as bending toward racial justice into a now "postracial" society. Meanwhile, tourists can indulge in Black suffering without consideration of extant structural racism. As Miles (2015, 70–71) summarizes well,

The LaLaurie story turns on crimes of racial violence, a kind of violence that our post–civil rights society rightly and vociferously rejects. In other, more mainstream social circumstances, listeners might feel self-conscious or even ashamed for enjoying the recounting of such brutal acts. However, the popularized LaLaurie tale includes within it an easy means for the dissolution of critical self-awareness. By participating in the castigation of a vicious slave mistress with whom they have nothing in common, tourists can distance themselves from her actions and relieve themselves of personal responsibility for indulging in scenes of abuse, racism, and torture.

LaLaurie's ghost story accomplishes much: it offers a thrilling tale, plays into accessible stereotypes of Black beastliness, abjectivity, and passivity, and sanitizes the New Orleans slave system by presenting LaLaurie as particularly brutal. Such a dynamic also occurs in the ghost story of Julie.

Julie, the "Tragic Mulatto"

In *Haunted New Orleans*, Troy Taylor (2010, 105–106) offers the story of Julie, a phantom woman who has been seen atop the roof of her home. As Taylor begins the story, Julie was a free "mixed" women of color, who were categorized as "mulattoes" or "octoroons." She met a wealthy white Frenchman at a plaçage party. Popular in the early 1800s, plaçage parties were held in ballrooms where white men and free women of color met. Since interracial marriage was outlawed, the women acted as their mistresses, and the men provided financial support and, in some cases, housing. Taylor (2010, 38) explains that, while some have viewed the women as "no better than prostitutes . . . [t]he girls were raised to be proper young women and were as well educated as the times allowed. They were free women and known for their beauty."

As Taylor (2010) continues in the story, Julie fell in love with a wealthy Frenchman. While he offered her material wealth, he could not provide what she truly desired: his love and hand in marriage. To prevent her from pressing the topic further, he challenged her, "If you love me you will get naked and go on the roof." It was mid-December and during a cold snap, so he did not expect her to follow through. Yet, after a night of drinking and card playing, he searched for her on the roof and found her dead. Now, Julie haunts the space of her demise, as Taylor (2010, 105–106) adds,

> They say that she walks on the rooftop of this building, completely naked and unprotected from the cold. As the wind slices around the eaves, the lovely phantom huddles in misery with her arms wrapped about her as if they can somehow shield her from the elements. The stories say that she huddles beneath the eaves throughout the night, only to vanish as dawn begins to color the sky. Those who have seen her, and have come to search for her in the night, will find no trace of her in the darkness, and yet the apparition will return to the rooftop the following night. She is doomed to repeat these actions over and over again—but only on the coldest nights of the year.

Like with LaLaurie's ghost story, Julie's tale does much to whitewash slavery. First, her story romanticizes interracial relationships within the context of New Orleans slavery. Such plaçage party narratives, as Laura Foley (2021, 10) puts it, "paint a picture of grand balls, a place where wealthy, well-educated women of mixed race would attend in hopes of becoming the mistress of a wealthy, white gentleman." A guide explained to us that the women's "parents wanted this because they were taken care of" by the white men. "They were given an allowance," she continued, "as well as food and clothing, and their children were sent to France for education." Yet, while such places most likely did exist, they did not resemble such grand balls (Aslakson, 2011). Often the men were not wealthy, the balls were not meeting places for plaçage relationships, mothers did not "bargain away their daughters," and many relationships were long term (Clark, 2015). They were, by and large, a mode of survival for many of these women (Foley, 2021, 17).

Tour guides and ghost books tell of a second theme in the story, that of romantic love. In *Ghost Stories of Old New Orleans*, Jeanne DeLavigne (1946, 30) describes Julie as such,

> But there was one slave, Julie, whose nature ran to dreams. Romance filled her strangely blended soul. She was an octoroon—seven streams of white blood, one stream of black. But her spirit cleared the black barrier at a single leap. She fell in love with her master—passionately, hopelessly, without reason or reserve.

One guide accounted for her desperation, that she went to Voodoo queen Marie Laveau for a love spell, and another affirmed that both her and the Frenchman are now in "eternal love in the afterlife." The depiction of Julie and the Frenchmen's unrequited love reinforces the "tragic mulatto" trope: Julie is unable to receive his affection, to clear the "black barrier at a single leap," and is doomed to a tragic end (Foley, 2021, 15). Her death, in turn, "becomes a punishment for the fact that she is not content with her 'place' in society" (21). Meanwhile, the Frenchman, like Louis LaLaurie, remains offstage, or is viewed in a sympathetic manner (Taylor, 2010, 109).

Like the brutalized enslaved, the screaming man, and Lisette who haunt LaLaurie's mansion, Julie eternally suffers within the story. Her ghost walks the rooftop naked "in misery with her arms wrapped about her . . . doomed to repeat these actions over and over again" (Taylor, 2010, 106). In their tour of a ballroom, Tiya Miles's (2015, 55) guide, Jeffries, spoke of two women of color who continue to haunt the building, one a matchmaker and the other an attendee. As she wrote in her notes,

> The center of the ballroom seemed to produce the highest number of energy spikes. Jeffries soon revealed the reason: black female ghosts from the nineteenth century were causing the disturbance. Jeffries explained that African American women of partial white "blood"—"quadroons, octoroons, and mulattos"—used to attend quadroon balls to meet their "protectors," wealthy white men who would take them as mistresses. These men might find true love with their mistresses and maintain two families, Jeffries said. She knew from her paranormal investigations that "two quadroons" haunt the ballroom today: one was a matchmaker of mistresses and their "protectors," and one was returning to the spot where she had "met her true love and been happiest."

These two ghosts, then, also join the ranks of those eternally trapped within the slaveholding society—while simultaneously representing rosy renditions of the balls.

Ultimately, Julie is put to work: her story reinforces a romanticized view of plaçage parties and interracial relationships, within the context of a seemingly racially progressive slave system. Her story also applies the tragic mulatto trope and white male blamelessness. In turn, she is rendered an object of punishment and thrill, with no recognition of the survival strategies employed by women participating in these engagements. In the final story, we turn to Marie Laveau, the New Orleans Voodoo queen.

Voodoo, Commercialization, and Demonization

On December 24, 2013, occult author Dorothy Morrison posted a message on Facebook bemoaning a recent experience she had in the city:

> As grand a time as we're having in New Orleans, we did make a rather disturbing discovery this morning: Someone painted Marie Laveau's tomb . . . I can hardly bring myself to say this . . . pastel PINK!!! WTF??? Someone local: Please find out who did this and make them change it back! It's disgusting, and I don't think Madame is very happy about it.

Marie Laveau's tomb was restored and unveiled on October 31, 2014 (Alvarado, 2020, 39–40), although, much to the city's chagrin, visitors continue to draw three X marks on her aboveground vault in hopes to be granted a wish, (Sircy, 2023, 41). Not just for a wish, but visitors go to her burial to potentially see her ghost (which is also said to haunt her home on St. Ann Street and was cited in the 1930s in a nearby drugstore (39)). Indeed, Laveau's tomb "is purportedly one of the most visited pilgrimage sites in the United States, second only to the King of Rock and Roll, Elvis Presley" (Alvarado, 2020, xi). Such attests to the power of Marie Laveau, a 19th-century free woman of color and Voodoo queen of New Orleans.

Voodoo, or Vodou or Vodun, originated in tribal religions of West Africa. Practitioners incorporate root work, distribute magical charms or "gris-gris," and believe in the power of spirits and charmed objects (Miles, 2015, 120–121). Voodoo traveled to New Orleans by enslaved Africans, largely through Haiti, with an estimated ten thousand arriving in 1809 alone. As Catholicism was the mandated religion through the Code Noir, practitioners blended Catholic saints with Voodoo deities to continue their practice, and they congregated at New Orleans's Congo Square on Sundays.

While a rich tradition grounded in slave struggle, Voodoo, historically and within the tourism industry, has been demonized and commercialized. In this section, we consider three ways Laveau, and Voodoo by proxy, is characterized in popular media and is marketed in the ghost tourism industry. The three marketed representations are of Laveau as a demonic figure, a manipulative spiritual entrepreneur, and an abolitionist saint. Each of these, we argue, is not consistent with the historical record of her life, produces one-dimensional characterizations of her, and minimizes a focus on broader forms of slave resistance and Voodoo's role within it.

Given Voodoo's connection to African survival and resistance, practitioners were readily demonized by white settlers. The *New Orleans Democrat* (1881) portrayed Laveau in her obituary as the leader of "that curious sect of super-stitious darkies who combined the hard traditions of African legends with the fetish worship of our creole negroes" (Long, 2007, xxv). The *Daily State Journal* (1873) described Laveau attending a court hearing "dressed in all the soiled frippery of an impecunious princess" whose "eyes had all the seeming of a demon that was dreaming" (Alvarado, 2020, xiv). As with the representation of Tituba in *The Crucible* (Miller, 1953), Robert Tallant (1946) also wrote of the Voodoo celebration, St. John's Eve, as replete with "snake worship," a "roaring bonfire," a "steaming cauldron," a "breast torn from a living chicken," and of drunken, naked "Voodoos" dancing in "spasmodic jerking and trembling" until they "fell to the ground unconscious," with others running into the lake in "wild screams of frenzy" (Long, 2007, 133).

Our tour guides rejected such demonization. One for instance, assured that Voodoo was not inherently demonic and compared it to Christianity. Both, he stated, "can have some bad actors," but, "for the most part, Voodoo is a peaceful religion." Yet, like the Hollywood witch of Salem, we find that New Orleans is also unable to let go of the demonic Voodoo image. Shops portray the dark aes-thetic of Voodoo when selling dolls, leggings, magnets, bottle openers, hot sauce, purses, key chains, and stickers. The names of tourist shops, too, appeal to the dark allure of Voodoo, such as Marie Laveau House of Voodoo, Reverend Zombie's House of Voodoo, Bloody Mary New Orleans Haunted Museum and Voodoo Pharmacy, and Voodoo Authentica (Foley, 2021, 59).[2]

Similarly, *American Horror Story: Coven*, which was often referenced by guides, reinforces the demonic representation of Laveau and, thus, Voodoo more generally. In one scene, Laveau, played by Angela Basset, reminds white witch, Fiona, played by Jessica Lange, that, "We're more than just pins and dolls and seeing the future in chicken parts." Yet, throughout the series, Laveau engages in animal sacrifice, raises the dead, and takes direction from the devil-ish spirit, Papa Legba. In the end, both Laveau and LaLaurie end up in hell, with Laveau meant to continuously dispense retributive justice on LaLaurie and her daughters for her brutality toward those she enslaved (O'Reilly, 2019).[3] Indeed, *American Horror Story*'s depiction thus echoes Tallant's (1946, 65–66) assertion that Laveau spoke with Lucifer. Thus , as Miles (2015, 199) concludes of the marketing of Voodoo, "Voodoo becomes the symbol for a dark spiritual plane, for secret lives, and for the subversive impulse."[4]

The second characterization of Laveau is as a manipulative spiritual entre-preneur. A common story in ghost books is that she placed three hot peppers beneath a judge's chair to assist in the acquittal of a wealthy young white man. The man is told to have subsequently offered her the residence on St. Ann Street (Nott, 1922; Taylor, 2010, 82–83). This is despite that the house was built for

Laveau's grandmother, Catherine Henry, after she purchased the lot in 1798 (Long, 2007, 60–61). Stories also tell of Laveau using her work as a hairdresser to learn secrets to manipulate local elites and white clientele (Long, 2007, 50). Through her spiritual manipulation, as Tallant (1946) argues, she became "the real boss of New Orleans,"

> [She could] have a policeman fired with one snap of her fingers and she could get one promoted with two snaps. . . . She just walked into a big politician's office and said "Do it! I is Marie Laveau and I wants it done." And he knew better 'n not to do. If he didn't something awful bad was sure gonna happen to him
>
> *(Long, 2007, 83).*

These accounts, nevertheless, appear to be false. Laveau had financial difficulties, was far from having the ability to sway local officials, and even her use of Voodoo could be overstated, with the chili pepper story being made up (Long, 2007). Either way, the notion of a mystical, all-powerful Black woman who manipulated her way through the secrets of Voodoo holds mainstream appeal.

The idea of Voodoo as a tool for individual betterment, too, caters to individual consumer marketing. That is, the *positive* aspects of Voodoo fix into an economy of spiritual goods that cater to individual self-growth. One can find in shops voodoo dolls, charms, stones, offering bags, oils, and spells for luck, love, financial gain, protection, and to win in court. A tourist may also visit one of the many fortune tellers sitting on Bourbon Street or in front of the St. Louis Cathedral Church. Miles (2015, 51–53) offers an example where, before embarking on a bus tour, passengers were given a gris-gris bag containing a penny, red beans, rice, a bay leaf, and a coupon for another tour. A gold tag affixed to the bag read, "Keep this bag with you at all times while in 'The most haunted city in America.'" Yet, without deeper exploration, the offering felt to Miles like a "cheap keepsake-cum-coupon that simplified as well as commercialized Voodoo (Vodou or Vodun) beliefs." To add, comparing Voodoo to Christianity, as one of our guides did, can omit historical context of African struggle—making it even easier to coopt for individual white consumer self-betterment (Woodland, 2023, 252).[5]

The third way Laveau is marketed is as a saint, an abolitionist, and a community leader (Long, 2007). *The Daily Picayune* (1881) described her as, "A woman with a wonderful history" who provided charity for the poor, nursed those with yellow fever and cholera, and had a knowledge of "the valuable healing qualities of indigenous herbs" (Long, 2007, xxiv). Numerous stories tell of her sitting with condemned prisoners and advocating for Louisiana to outlaw public hangings (Tallant, 1946; Long, 2007, 153). Many also claim her residence was a safehouse for the Underground Railroad (Long, 2007, 72), given, for instance, that she maintained a statue of St. Marron, the patron saint of runaway slaves (Alvarado, 2020, 12). Scholars have also contended that Laveau, with her second husband, Christophe Glapion, a white man, "bought slaves in order to liberate them" and offered charms "to protect them on their journey north to liberty in Canada" (Long, 2007, 72; see Fandrich, 2005). Such a saintly image was presented by our guides, with one stating that she "is a woman of color who demands all the respect for her practices."

Laveau certainly did support others, by, for instance, paying the bond for an incarcerated woman of color and by educating an orphan boy (Long, 2007, 210). Yet, she also owned slaves with no intention of freeing them, as many people of color did in New Orleans, including her father, Charles Laveau (46). Between 1828 and 1854, Laveau and Glapion bought and sold eight slaves: Eliza, Molly and her sons Richard and Louis, Peter, Irma and her son Armand, and Juliette (72). At the same time, she did face historical and collective trauma as a descendent of enslaved people and from being a woman of color in a racially stratified society—with any secrets and manipulation perhaps being a matter of survival (Fandrich, 2005, 11; Alvarado, 2020, 4). Nevertheless, her public activism appears overstated. Her folk hero image, too, can present her as the strong, assertive Black woman which, as bell hooks (2014) has warned, tends to elide the pain and vulnerability of Black women (Thompson-Spires, 2022, 117).

Overall, these three ways of marketing Laveau, as demonic, as spiritual entrepreneur, and as abolitionist saint, do not capture the more complex historical aspects of her life. Like LaLaurie and Julie, as well as Tatebe in the previous chapter, ghost stories tend to speak more to racial ideology than historical documentation. As Carolyn Long (2007, 207) concludes, "As with all legendary villains, heroes, and saints, we have created an imagined Marie Laveau in fulfillment of our own fears and desires," as she morphs between "saintly provider to she-devil mother goddess." In turn, stories surrounding LaLaurie, Julie, and Laveau rely on racist stereotypes to whitewash the brutalities of slavery. They also do not discuss collective resistance. How might we see the marks of slavery in New Orleans without relying on ghost stories of individual brutality, romantic tragedy, or individual good will?

Spectral Markers

One need not stick with fantastically brutal stories of Black suffering to witness the marks of slavery in New Orleans. We noted when eating lunch at the Original Pierre Maspero's restaurant two poles separating the room. It was where enslaved people were divided by gender for sale. As part of the restaurant's logo, pictured on top of the menu is the French fleur-de-lis symbol, which was branded on enslaved people who attempted escape. Outside, there is a marker reading, "Within this historic structure slaves were sold." Outside of the restaurant, we came across another marker, now for the St. Louis Exchange Hotel, which states, "auctioneers sold off land and goods as well as thousands of enslaved people." These markers, while perhaps blending into the environment, are key to telling the story of slavery in New Orleans. Indeed, to aid in the seeking of those enslaved in the city, in 2018, the city unveiled the New Orleans Slave Marker Tour App, which is a smartphone application that provides an overview of each site and offers "first-person testimonies of enslaved individuals who were bought and sold in the New Orleans Market" (New Orleans Slave Trade).

Monuments of slave owners and traders also dot the land. There is Andrew Jackson Square, where enslaved people were tortured and executed. In the center is a statue of Major General Andrew Jackson of the U.S. Army who, in 1815, led a small army to victory against British troops at the Battle of New Orleans. As a monument, it also commemorates the same man who signed the 1830 Indian Removal Act, which initiated the Trail of Tears. It should be noted that one guide, a young white man, also critiqued the Jackson monument for this history.[6] In the Metairie Cemetery, where a tour guide drove us in her van, we also came across the Army of Tennessee, Louisiana Division monument which honors the soldiers who fought for the Confederate Army of Tennessee. A bronze statue depicts Confederate General Albert Sydney Johnston riding his horse, Fire Eater, into the battle of Shiloh in 1862 (Figure 2.2). In these monuments, we can see a valorization of men, and white men in particular, who did much to contribute and sustain the slave and colonial system.

Slave narratives also give voice to the horrors of the slave trade from those who experienced it—revealing every day, or horrifically ordinary, violence. In

FIGURE 2.2 Army of Tennessee, Louisiana Division Monument

Slave Life in Georgia, John Brown (1855, 112–118) describes the New Orleans marketplace in detail, of being caged, "plumped up," and made to dance for sale (see Turner, 2022):

> The slaves are brought from all parts, are of all sorts, sizes, and ages, and arrive in various states of fatigue and condition; but they soon improve in their looks, as they are regularly fed, and have plenty to eat. As soon as we were roused in the morning, there was a general washing, and combing, and shaving, pulling out of grey hairs, and dyeing the hair of those who were too grey to be plucked without making them bald. When this was over—and it was no light business—we used to breakfast, getting bread, and bacon, and coffee, of which a sufficiency was given to us, and that we might plump up and become sleek. Bob would then proceed to instruct us how to show ourselves off.
>
> The buying commenced at about ten in the morning, and lasted till one, during which time we were obliged to be sitting in our respective companies, ready for inspection.
>
> After dinner we were compelled to walk, and dance, and kick about in the yard for exercise; and Bob, who had a fiddle, used to play up jigs for us to dance to. If we did not dance to his fiddle, we used to have to do so to his whip, so no wonder we used our legs handsomely, though the music was none of the best
>
> (*Hartman, 2022, 59–60*).

Enslaved women also spoke of their experiences in the market, as documented by Saidiya Hartman (2022, 59):

> Millie Simpkins stated that before they were sold, they had to take all their clothes off, although she refused to take hers off, and roll around to prove that they were physically fit and without broken bones or sores. Usually any reluctance or refusal to disrobe was met with the whip. When Mattie Gilmore's sister Rachel was sold, she was made to pull off her clothes. Mattie remembered crying until she could cry no more, although her tears were useless.

As this passage attests, those enslaved such as Simpkins also resisted by refusing to disrobe. Others threatened self-harm and the taking of their own lives (Johnson, 1999).

It is also important to draw on the role of Voodoo (Vodou or Vodun) in Black resistance, as it is so often demonized and commercialized, as evident in the ghost tourism industry. For enslaved people Voodoo acted as a "lifeline," as a cultural and spiritual resource stretching to the African Yoruba religion (Miles, 2015, 122). The Yoruba religion, as Tracey Hucks (2012, 11, xiii, xix) writes, "invokes a meaningful connection to Africa as 'originary space' that substantiates human value and provides restorative ontological, historical, and spiritual integrity." Within chattel slavery, as Miles (2015, 121) further explains, belief "in a rich and potent African-inspired spirit world sustained black identity, strength, and will"; it presented "slaves as human beings worthy of dignified life and invested with spiritual efficacy."

Voodoo also had revolutionary influence. On August 22, 1791, in the forests of the Morne Rouge of then French colony Saint-Domingue, High Priest Dutty Boukman led those enslaved into the rebellion that would spark the 1791–1804 Haitian Revolution, which dismantled slavery and colonial rule in the French colony. Indeed, by the late 1780s, French Caribbean colonies made up two-fifths, or 40 percent, of the Western world's sugar and coffee production, with the French government increasingly dependent on colonial commerce (Peabody, 2002, 3). Before the rebellion, Boukman offered a Vodou ritual: he provided instructions, led incantations, initiated the sucking of the blood of a pig, and spoke a prayer in Creole:

> The god who created the sun which gives us light, who rouses the waves and rules the storm, though hidden in the clouds, he watches us. He sees all that the white man does. The god of the white man inspires him with crime, but our god calls upon us to do good works. Our god who is good to us orders us to revenge our wrongs. He will direct our arms and aid us. Throw away the symbol of the god of the whites who has so often caused us to weep, and listen to the voice of liberty, which speaks in the hearts of us all
>
> *(James, 1989, 86–88).*

The Haitian Revolution sparked fear in white planters. To break collective and revolutionary spirit in New Orleans, police infiltrated Voodoo rituals by enforcing the slave code banning "unlawful assembly of slaves and free persons." They arrested, jailed, fined, and flogged participants (Long, 2007, 128). And yet Voodoo—and Black resistance more generally—prevailed. In the summer of 1850, a much less-known figure, Voodoo priestess Besty Toledano, "went to court to protest the persecution of those engaged in" Voodoo ceremonies (210).

Planters' fears came to fruition in 1811, with the biggest slave revolt in American history. The German Uprising began at the Manual Andry plantation, 41 miles northwest of New Orleans. After months of conspiring, two hundred to five hundred enslaved men marched from the German Coast sugar plantations to New Orleans (Genovese, 1976). They adopted code language and wielded clubs, farm tools, and knives (Smith, 2022, 175). They burned five plantation houses, several sugar houses, as well as crops, and they killed two white men before being hunted down by white militias and federal troops. The retaliating force executed and decapitated "more than one hundred of the rebels, whose heads they publicly displayed in an effort to intimidate and discourage further revolts" (Rasmussen, 2011; Alvarado, 2020, 10). Nevertheless, such action shows that those enslaved were not docile, animalistic, voiceless masses but engaged in spirited action against white supremacy.

Indeed, some guides did offer more critical interpretations of slavery in New Orleans and how it echoes today. A young Black woman tour guide offered a critical reading of the LaLaurie story. She limited the description of those brutalized and prefaced by stating the attic scene was "the most difficult part," thus offering gravity to the account. She also identified that the fire was set by a Haitian cook who "would rather die than live in the house" and reminded that her actions had set free those trapped in the attic (the woman may have been Arnante, who was a cook, laundress, and domestic [Long, 2012, 101]). She also reminded us that LaLaurie's husband was, after all, a doctor and that his treatment toward those enslaved is evident in Black medical harm today. Black people, she added, "are viewed as being immune to pain," and "Black women die at a higher rate when giving birth" (see Washington, 2008). "When I go to the hospital," she shared, "I am just prescribed Tylenol." She thus presented how, as Kidada Williams (2023, xxv) puts it, the "present and past are chaotically and violently fused."[7]

Such critical telling is key to draw connections between New Orleans slavery as it haunts the present—including the current socioeconomic conditions of the city. As Laura Foley (2021, 42) reminds,

[T]he groups that are highlighted within the ghost stories found in New Orleans reflect living, breathing marginalized communities who are still dealing with repercussions from their historical struggle today.

This is apparent with, as a final note, the ramifications of category-four Hurricane Katrina, where 18,000 people died and 80 bodies were not claimed. We visited the Hurricane Katrina Memorial on a tour. As the spot was once used by Charity Hospital to bury poor and indigent people, our guide claimed it had high spiritual activity. In this, we can further see how the city is haunted by the racial violence that occurred in its wake and aftermath.

Many were not able to evacuate the city; this mostly included the impoverished, which were disproportionately Black residents. Conditions in the Superdome were poor, and residents felt as though they were occupied by police and military presence (Page, 2022, 116). Some white residents even armed themselves and hunted displaced Black people (Hayes and Kaba, 2023, 53). Yet, there was still care and community (Page, 2022). This included women, and Black women in particular, who eased "the hunger and thirst of babies and toddlers left in their care in the sweltering heat and the inhumane conditions associated with post-disaster survival" (Jones-DeWeever, 2008). To be sure, these women, too, have been ignored in the media, affording what Deborah Douglas (2022, 377) describes as depresencing, which "never acknowledges presence at all. When deployed, people just look right through Black women as if they weren't there."

Within the tourism industry, we can create presence—by focusing on markers and monuments; on revolution and rebellion; on disaster and racial poverty; and on collective care, led by Black women. In turn, we see the haunting of unresolved social violence, as it manifests materially and ideologically, as well as the spirit of collective care and revolt.

Conclusion—From Stereotypes to Spectral Markers

Yet, the history of slavery and revolt is largely whitewashed in the New Orleans ghost tourism industry: LaLaurie is presented as an exceptionally immoral slave owner; Julie is a tragic mulatto, doomed to eternally haunt the roof of her demise; and Laveau is viewed as a demonic, manipulative, or saintly figure. Yet, as with Salem, the haunting horrors of unresolved social violence in the city can be seen in markers, monuments, movement, and by critical tour guides. In this regard, we encourage tourists to look between the cracks of the city, and of dominant ghost stories, to find spectral markers of unresolved social violence. And with this, we move outside of New Orleans to the Myrtles Plantation, "One of the most haunted homes located in Louisiana."

Notes

1 LaLaurie escaped to France and died in Paris, although others suggest she relocated to Baton Rouge, Louisiana. Some have rumored that she was killed by a wild boar on a hunting expedition (Cable, 1889), although there is no evidence for this (Long, 2012, 134).
2 Even the term "dark" in "dark tourism" can connote Black otherness in the sense that "light" was associated by Enlightenment thinkers such as Robert Boyle with whiteness and white skin. Isaac Newton wrote, as relayed by Ibram X. Kendi (2016, 45),

> "Whiteness is produced by the Convention of all Colors," he wrote. Newton created a color wheel to illustrate his thesis. "The center" was "white of the first order," and all the other colors were positioned in relation to their "distance from Whiteness." In one of the foundational books of the upcoming European intellectual renaissance, Newton imagined "perfect whiteness."

3 Of note, guides often spoke of the miscasting of LaLaurie in *American Horror Story: Coven*, stating she was younger than Kathy Bates. Yet, they did not discuss the mischaracterization of Laveau.
4 The Indian burial ground horror trope was also adopted in a tour that Tiya Miles (2015, 124) attended. As she writes, "In the Haunted History of New Orleans tour, the guide proclaimed that cannibal Indians had performed their sacred rituals in the spot where the city's deadly fires had later erupted."
5 Not only Voodoo dolls, but one haunted museum had Black Americana items, including an ashtray with the mammy figure on it.
6 It should be reminded that tour guides are not monolithic and can at times challenge dominant historical narratives, as we will discuss further in the latter part of this section.
7 New Orleans has always had a substantial Black population, with guides and visitors reflecting a much more racially diverse population than in Salem or Gettysburg. As of 2022, the Black population was 55 percent of the total population (Census Reporter, 2022).

References

Abrams, Eve. 2015, June 15. "Remembering New Orleans' overlooked ties to slavery." *NPR*. https://www.npr.org.

Alvarado, Denise. 2020. *The Magic of Marie Laveau: Embracing the Spiritual Legacy of the Voodoo Queen of New Orleans*. Boston: Weiser Books.

American Horror Story: Coven. Season 3, episode no. 1, "Bitchcraft," first broadcast October 9, 2013 by FX. Directed by Alfonso Gomez-Rejon.

Aslakson, Keith. 2011. "The 'Quadroon-Plaçage' Myth of Antebellum New Orleans: Anglo American (Mis)Interpretations of a French-Caribbean Phenomenon." *Journal of Social History*, 45(3): 709–734.

Buncombe, Andrew. 2015. "Hurricane Katrina 10th Anniversary: New Orleans Is Haunted by the Death of Vera Smith." *Independent*. https://www.independent.co.uk.

Cable, George Washington. 1889. *Strange True Stories of Louisiana*. New York: Charles Scribner's Sons Publisher.

Census Reporter. 2022. New Orleans, LA. https://censusreporter.org.

Chamerovzow, Louis A. 1855. *Slave Life in Georgia: Narrative of the Life, Sufferings, and Escape of John Brown*. London: W. M. Watts.

Chavez, Roby. 2022. "New Orleans Was Once the Center of U.S. Slave Trade. This Artist Wants to Make Sure We Don't Forget." PBS, https://www.pbs.org.

Clark, Emily. 2015. *The Strange History of the American Quadroon: Free Women of Color in the Revolutionary Atlantic World*. Chapel Hill: The University of North Carolina Press.

Daily Picayune. 1881, June 17. "Death of Marie Laveau—A Woman with a Wonderful History, Almost a Century Old, Carried to the Tomb Yesterday Evening," 8.

Daily State Journal. 1873, April 16. "No title."

DeLavigne, Jeanne. 1946. *Ghost Stories of Old New Orleans*. New York: Rinehart & Company.

DeRosa, Robin. 2009. *The Making of Salem: The Witch Trials in History, Fiction and Tourism*. Jefferson: McFarland.

Dickey, Colin. 2016. *Ghostland: An American History in Haunted Places*. Westminster: Penguin.

Douglas, Deborah. 2022. "Hurricane Katrina." In *Four Hundred Souls: A Community History of African America, 1619–2019*, edited by Ibram X. Kendi and Keisha N. Blain, 374–377. New York: One World.

Fandrich, Ina J. 2005. "The Birth of New Orleans' Voodoo Queen: A Long-Held Mystery Resolved." *Louisiana History*, 293–309.

Foley, Laura. 2021. "The Haunted History of New Orleans: An Exploration of the Intersectionality between Dark Tourism, Black History, and Public History." *ProQuest Dissertation Publishing*. Glassboro: Rowan University.

Follett, Richard J. 2005. *The Sugar Masters: Planters and Slaves in Louisiana's Cane World, 1820–1860*. Louisiana State University Press.

Genovese, Eugene D. 1976. *Roll, Jordan, Roll: The World the Slaves Made*. New York: Vintage.

Getting Curious with Jonathan Van Ness (2023). "How Did New Orleans Become New Orleans (Part 1)."

Gilmore, David D. 2012. *Monsters: Evil Beings, Mythical Beasts, and All Manner of Imaginary Terrors*. Philadelphia: University of Pennsylvania Press.

Hartman, Saidiya. 2022. *Scenes of Subjection: Terror, Slavery, and Self-Making in Nineteenth-Century America*. New York: W.W. Norton & Company.

Hayes, Kelly, and Mariame Kaba. 2023. *Let This Radicalize You: Organizing and the Revolution of Reciprocal Care*. Chicago: Haymarket Books.

Historic New Orleans Collection. "New Orleans, Slave Market of the South." https://www.hnoc.org.

hooks, bell. 2014. *Ain't I a Woman: Black Women and Feminism*. Abington-on-Thames: Routledge.

Hucks, Tracey E. 2012. *Yoruba Traditions and African American Religious Nationalism*. Albuquerque: University of New Mexico Press.

James, C.L.R. 1989. *The Black Jacobins: Toussaint L'Ouverture and the San Domingo Revolution*. New York: Vintage.

Johnson, Walter. 1999. *Soul by Soul: Life Inside the Antebellum Slave Market*. Cambridge: Harvard University Press.

Jones-DeWeever, Avis. 2008, April 1. "Women in the Wake of the Storm: Examining the Post-Katrina Realities of the Women of New Orleans and the Gulf Coast." *Institute for Women's Policy Research*. https://iwpr.org.

Kendi, Ibram X. 2016. *Stamped from the Beginning: The Definitive History of Racist Ideas in America*. Boston: Bold Type Books.

King, Amy K. 2017. "A Monstrous (ly-Feminine) Whiteness: Gender, Genre, and the Abject Horror of the Past in *American Horror Story: Coven*." *Women's Studies*, 46(6): 557–573.

Long, Carolyn Morrow. 2007. *A New Orleans Voudou Priestess: The Legend and Reality of Marie Laveau*. Gainesville: University Press of Florida.

Long, Carolyn Morrow. 2012. *Madame LaLaurie, Mistress of the Haunted House*. Gainesville: University Press of Florida.

Louisiana Office of Public Health. 1934. "Infectious Disease Epidemiology." *Annual Report*.

Love, Victoria Cosner, and Lorelei Shannon. 2011. *Mad Madame LaLaurie: New Orleans' Most Famous Murderess Revealed*. Mount Pleasant: Arcadia Publishing.

Martineau, Harriet. 1838. *Retrospect of Western Travel*. London: Saunders & Otley.

Miles, Tiya. 2015. *Tales from the Haunted South: Dark Tourism and Memories of Slavery from the Civil War Era*. Chapel Hill: University of North Carolina Press.

New Orleans Democrat. 1881, June 17. "Marie Lavaux—Death of the Queen of the Voudous," 8.

New Orleans Slave Trade. "New Orleans Slave Trade Marker Tour & Audio Guide—Step into History." https://www.neworleansslavetrade.org/new-page.

Nott, William G. 1922. *Times-Picayune*.

Olivarius, Kathryn Meyer McAllister. 2022. *Necropolis: Disease, Power, and Capitalism in the Cotton Kingdom*. Cambridge and London: Harvard University Press.

O'Reilly, Jennifer. 2019. "'We're More Than Just Pins and Dolls and Seeing the Future in Chicken Parts': Race, Magic and Religion in *American Horror Story: Coven*." *European Journal of American Culture*, 38(1): 29–41.

Page, Cara. 2022. "Spiritual Conditions: Mapping the Origins of Healing Justice." In *Healing Justice Lineages: Dreaming at the Crossroads of Liberation, Collective Care, and Safety*, edited by Cara Page and Erica Woodland, 110–118. Berkeley: North Atlantic Books.

Peabody, Sue. 2002. *"There Are No Slaves in France" The Political Culture of Race and Slavery in the Ancien Régime*. Oxford: Oxford University Press.

Prison Policy Initiative (PPI). "Louisiana Profile." https://www.prisonpolicy.org.

Ralph, Laurence. 2022. "The Code Noir." In *Four Hundred Souls: A Community History of African America, 1619–2019*, edited by Ibram X. Kendi and Keisha N. Blain, 57–61. New York: One World.

Rasmussen, Daniel. 2011. *American Uprising: The Untold Story of America's Largest Slave Revolt*. New York: HarperCollins Publishers.

Sillery, Barbara. 2001. *The Haunting of Louisiana*. New Orleans: Pelican Publishing.

Sircy, Allen, 2023. *Southern Ghost Stories: New Orleans*

Smith, Clint. 2022. "The Louisiana Rebellion." In *Four Hundred Souls: A Community History of African America, 1619–2019*, edited by Ibram X. Kendi and Keisha N. Blain, 173–176. New York: One World.

Smith, Kalila. 2016. *New Orleans Ghosts, Voodoo and Vampires* (7th ed.). New Orleans: De Simonin Publications.

Stowe, Harriet Beecher. 1852. *Uncle Tom's Cabin*. Boston: John P. Jewett & Company.

Tallant, Robert. 1984. *The Voodoo Queen*. New Orleans: Pelican Publishing.

Taylor, Troy. 2010. *Haunted New Orleans: History & Hauntings of the Crescent City*. Mount Pleasant: Arcadia Publishing.

Thompson-Spires, Nafissa. 2022. "Lucy Terry Prince." In *Four Hundred Souls: A Community History of African America, 1619–2019*, edited by Ibram X. Kendi and Keisha N. Blain, 115–118. New York: One World.

Turner, Sasha. 2022. "The Slave Market." In *Four Hundred Souls: A Community History of African America, 1619–2019*, edited by Ibram X. Kendi and Keisha N. Blain, 85–88. New York: One World.

Washington, Harriet A. 2008. *Medical Apartheid: The Dark History of Medical Experimentation on Black Americans from Colonial Times to the Present*. New York: Doubleday Books.

Williams, Kidada E. 2023. *I Saw Death Coming: A History of Terror and Survival in the War against Reconstruction*. New York: Bloomsbury Publishing.

Woodland, Erica. 2023. "Co-Optation, Critical Questions, and Contradictions: The Future of Healing Justice." In *Healing Justice Lineages: Dreaming at the Crossroads of Liberation, Collective Care, and Safety*, edited by Cara Page and Erica Woodland, 251–264. Berkeley: North Atlantic Books.

3

ESCAPE FROM THE MYRTLES PLANTATION

Introduction

We drive from New Orleans on Interstate 10 west to St. Francisville, Louisiana. There is a notable contrast from the French Creole architecture of New Orleans as we traverse the Atchafalaya Basin, or Swamp Freeway. We are on the Mississippi's Yazoo Delta, a region settled in the 1830s by wealthy planters who cleared the swampy, snake-infested land to cultivate cotton in the fertile soil (Duncan, 2016). We make our way into St. Francisville, which sits north of Baton Rouge, and make it to the Myrtles Plantation. The plantation is a bed and breakfast and, according to promotional material, is "[o]ne of the most haunted homes located in Louisiana."

The Myrtles's story begins in 1798 with General David Bradford.[1] In 1794, Bradford fled to the then-Spanish colony Bayou Sara to avoid arrest in Pennsylvania for not complying with the federal whiskey tax, granting him the monicker "Whiskey Dave." In 1796, he obtained a land grant of 650 acres from the Baron de Carondelet and, a year later, completed the home and called it "Laurel Grove." After his death in 1808, his wife, Elizabeth, acquired the property. In 1817, their daughter, Sarah Matilda Bradford, married Clark Woodruff, who later gained ownership. Sarah and their two children, James and Cornelia Gale, died between 1823 and 1824. In 1834, Woodruff sold the property to Ruffin Gray Stirling who, with his wife, Mary Catherine Cobb, enlarged the grounds to over five thousand acres and remodeled the two-story mansion. They planted crepe myrtle trees and renamed the plantation The Myrtles. The Stirling family sold the property in 1889, and, after numerous transactions, Marjorie Munson purchased it in 1950 (see, Miles, 2015, 87; Kermeen, 2007, 322–323).

DOI: 10.4324/9781003397809-4

Ghosts began to appear after Munson's purchase, as she reported hearing "apparitions and disembodied voices in the house" (Newman, 2021).[2] The phantom presences, however, would not become a marketable mainstay until the property was turned into a bed and breakfast in 1980 by James and Frances Myers. In *The Myrtles Plantation* (2007), Frances Kermeen (formally "Myers") recounts a number of ghosts at the Myrtles, including that of Chloe. Kermeen describes her as "a mulatto slave dressed in green frock with a green turban wrapped around her head" (142). In 1993, John and Teeta Moss bought the property and continued marketing the bed and breakfast as a haunted excursion, with Chloe being one of 13 ghosts they report on the grounds. Each year, 40,000 visitors come from around the United States, many hoping to experience one of these ghosts firsthand (Miles, 2015, 86). The plantation has been featured on *Travel Channel* (2005), *National Geographic* (2008), and *HGTV* (2014).

How are ghosts, such as Chloe, marketed on the plantation? How is plantation life portrayed more generally? To explore these questions, we spent a day at the Myrtles. We went on a day, night, and self-guided tour. We walked the grounds and went to the onsite restaurant and gift shop. On top of this, we considered news media and website material. In this chapter, we identify two prominent ways the Myrtles Plantation is marketed. First is through the haunting narrative, which tells of 13 ghosts that inhabit the property. Stories of enslaved ghost Chloe, in particular, reinforce anti-Black stereotypes, such as the Jezebel and mammy tropes, and they commercialize Black suffering. Second is the antebellum splendor narrative, which portrays the bucolic life of antebellum living. In resisting this whitewashed portrayal of slavery, we revisit Chloe's story, like we did with Tatebe's in Salem, to critically engage with slavery, resistance, and how the legacy of slavery persists today.

"The Legend of Chloe"

While the bed and breakfast boasts 13 ghosts, none of them is more popular than Chloe. According to promotional material surrounding Chloe, after purchasing the property in 1992, the Mosses snapped photos for insurance purposes. They found a shadowy figure in one picture: it was of a woman who appeared to be wearing a turban. They brought the photo to a patent researcher who

> discovered that all of the physical measurements of the apparition were of human dimensions and proportions. . . . The circumference of the head, the length of the shoulder to the elbow and the length of the elbow to the wrist were all indicative of a human.

The Mosses determined that it must be Chloe, referring to her in promotional materials as "a slave girl standing between two of the buildings on the plantation," that of the mansion and of the General's Store. They subsequently marketed the photo as a postcard, deeming it "the Chloe postcard."

While Chloe was introduced in Kermeen's (2007) *The Myrtles Plantation*, she would soon gain a larger backstory. As *Unsolved Mysteries* (2001) tells it, Chloe was "caught eavesdropping on the judge." For punishment, he had her ear cut off. As she was "fearful of also being sent to the fields to do hard labor, Chloe made a plan." She "decided to grind poisonous oleander leaves into a birthday cake, hoping to make the family ill, then she would endear herself by nursing them back to health." Yet, her scheme backfired, "Young Cornelia, her mother, and sister all died from the poison." It is believed, the episode continues, that "she was killed by a mixed mob" of "black and whites." "Of course, the other slaves," the program adds, "were probably afraid of what was going to happen to them." Chloe was "supposedly beaten, and then thrown into the river." Subsequently, the show continues, "It is said that Cornelia, along with her mother and sister, have joined Chloe in eternally haunting the house."

Rebecca Pittman (2016, 97) offers a similar rendition of Chloe's ghost story. Upon eavesdropping, Chloe realized that her placement in the household was in jeopardy. She hatched a plan to bake a birthday cake for the ninth birthday of Woodruff's daughter, Mary Octavia, and mix it with "some poisonous crushed oleander leaves." Utilizing "root" medicine she learned from her grandmother, Chloe hoped "to nurse [his daughters] back to health." As Pittman writes,

> On the morning of October 3rd, Chloe assembled her bowls and ingredients for the cake she was making for the little girl's birthday party. The oleander springs were hidden in a box on a shelf beneath the pump-handle sink, where she had placed them for the day before. Her hands shook as she blended together the flour, milk and eggs for the batter. When the kitchen was empty of other slaves for a few minutes, she hurriedly heated a pot of water, and dumped in the red flowers and leaves from the poisonous oleander plant. Hurriedly, she stirred the boiling mixture, constantly looking over her shoulder for prying eyes. It gave off a sweet, earth odor she had not accounted for.

Unfortunately, she misjudged the amount of oleander, and the mixture "caused the children to die." Pittman (2016, 100) adds that the "kitchen slaves, fearful for the safety of the slave community," told the "field hands" of the mishap:

> Beneath a watchful moon, the men tied the rope around a large oak branch. The other end was looped around the neck of the screaming girl and pulled taut. They hoisted her into the air. After several minutes, she was cut down and carried along the slope to the shore of the Mississippi. They watched as her small form disappeared downstream.

In *The Haunting of Louisiana*, Barbara Sillery (2001, 19) also reports Chloe's violent death when concluding the tale, as based on Teeta Moss's rendition of the story,

Chloe in a panic told the other slaves what she had done because she really did not intend to kill Sara and the girls. Out of respect and fear of the Judge, Chloe's fellow servants dragged Chloe outside, and hung her from a large oak tree on the property.

National Geographic (2008), too, explains that "fearing that they would be accused of murder by association, Chloe's fellow slaves dragged her from bed that night, hanged her, then threw her body in the river."

The basic tenants of this story were told to us on both of our tours of the mansion. Standing in the entryway, underneath the French chandelier and next to a haunted mirror and piano, our guide offered Chloe's story (see Potter, 2016). Like the examples cited earlier, she included a variety of themes worthy of critical exploration. As she told it,

This is where history and mystery collide. Chloe worked in the home and was adored by the children. The wife was jealous. Chloe had a nasty habit of eavesdropping, as she knew how to keep her place with information. She was not good at it but would only receive a slap on the wrist. However, one day she was listening in on prominent men in the gentlemen's room. Woodruff had her ear cut off, as you lost the appendage you had committed the crime with. She was banished to the kitchen, which would have been extremely hot given the Louisiana weather.

Chloe planned to poison a birthday cake with oleander to secure her place. [The guide introduces us to a plant, joking that it was not really poisonous.] Chloe was naïve and added two handfuls. Later on, she did not hear the children call her sick, so she panicked. The wife and two children were dead, as the baby did not take any of the cake since they were on "mother's milk." Knowing those enslaved would be punished, they tried her for murder, hung her body, and dumped it into the Mississippi River.

It did not work. She comes back and is one of the active 13 ghosts. Woodruff moved, as he couldn't stand being there anymore. He planted the myrtle trees in memory of his lost family.

It is notable that there is no record of the Woodruffs owning any enslaved people named Chloe. Mary Octavia lived into adulthood, and "Sara, James, and Cornelia Woodruff were not killed by poisoning, but instead succumbed to yellow fever" (Moses, 1999). Yet, Chloe remains put to work on the plantation. As an enslaved person—or a representation of those enslaved (and of Black people more generally)—she is, according to Saidiya Hartman (2022, 8), an imaginative surface where racist ideas are projected. The captive body, thus, is yoked "to the ambitions, whims, fantasies, and exploits of the owner" (82) and is therefore made to "speak the master's truth" (59). In this regard, the "slave" is fungible in the sense that "they can be anything they are needed or desired to be. They are an abstract, replaceable, and interchangeable object, subject to the material and psychic appetites of those who own them" (Petersen, 2024, 13). In Chloe's story, guides and media programming thus project onto her a series of controlling images of Black women (Collins, 2008), which persist in the past and present. These include framing her as Jezebel, mammy, childlike, and conniving.

To the Jezebel trope, Chloe has been described as a mistress (Moses, 1999) and as a "household servant" having an "affair" with Woodruff (*National Geographic*, 2008). Such phrasing highlights her promiscuity, while down-playing the power imbalance between her and Woodruff. The Jezebel trope, accordingly, portrays Black women as "manipulative sexual temptresses who brought on and deserved their fate" (Miles, 2015, 94). The sexualization of African women is apparent in early racist texts. In his 1684 taxonomy of racial classification, French physician and travel writer François Bernier described African woman as follows, "Those cherry-red lips, those ivory teeth, those large lively eyes . . . that bosom and the rest . . . I dare say there is no more delightful spectacle in the world" (Kendi, 2016, 56). French com-mercial agent Jean Barbot also wrote in 1732 that African women possessed a "temper hot and lascivious, making no scruple to prostitute themselves to Europeans for a very slender profit, so great is their inclination to white men" (Kendi, 2016, 43). To be sure, the fetishizing of African woman by white scholars, artists, and travelers was born on a mixture of the grotesque, exotic, and scientific,[3] all which represent Black women Otherness.[4]

Chloe is simultaneously portrayed through the mammy trope, as the faithful, obedient domestic servant (Collins, 2008, 80). Kermeen (2007, 143) writes of a guest describing Chloe as "a large, homely woman with a very square jaw," and the former owner relayed to her that Chloe "goes from room to room, carrying her night-candle, checking to be sure that everyone is safe and warm" (68; Miles, 2015, 97). Such depictions imagine unsexed Black women as serving "the white family rather than undermining it" (Miles, 2015, 97; see McElya, 2007). As Patricia Collins (2008, 80) describes of the trope,

> By loving, nurturing, and caring for her White children and "family" better than her own, the mammy symbolizes the dominant group's perceptions of the ideal Black female relationship to elite White male power.

Yet, at the same time, Chloe is also infantilized. We noted this when touring her "favorite room," the French furniture room.

> The room includes a bed, a Victorian dress divider, and a chair in the corner. After showing us ghostly handprints on the bed, the guide points to a rocking chair. She explains that it is Chloe's favorite chair. She directs us to stand by it, implying that something might happen as we get close. She then shows us Chloe's favorite doll, named Charlotte.

Indeed, a guide told Miles (2015, 90) that Chloe was 13- or 14-years-old. Chloe also appears childlike in her naivete when adding too much oleander into the cake. While her age was never mentioned, such descriptions fit the more general view of enslaved people as childlike, which justifies the paternalism of the master-slave relationship (Johnson, 2001b).

Chloe is also presented as conniving and thieving. In *The Haunting of Louisiana*, Barbara Sillery (2001) refers to her as "conniving Chloe," as she steals earrings to adorn her single ear (Miles, 2015, 96). In the main entrance, our guide also informed us that Chloe likes to steal earrings from visitors for her left ear and showed us a case containing the jewelry (Figure 3.1). She also described Chloe as having a "nasty habit of eavesdropping." Similarly, in her tour, Holly Ann Vaughn (2012, 37) identified guides referring to Chloe as "nosey," a "gossip," and even "uppity." She wrote in her notes,

> [O]ne guide even used the word "uppity." (The use of the term was somewhat ambiguous; the way the guide embedded the term in her story somehow managed to suggest that Chloe was both sassy to her white owners and displayed an attitude of superiority toward fellow slaves—a liminal position indeed.) The stories and their purveyors would have us believe that she eavesdropped on the goings on in the house and reported back to

FIGURE 3.1 Case of Chloe's "Stolen" Earrings

the other slaves. Her primary objective—according to some tellings—was to improve her position with the "family." Her primary objective—according to others—was to demonstrate her superior status in relation to the other slaves.

While not evident in our visit, guides have also employed the demonic Voodoo theme to heighten the thrilling experience of the mansion. Indeed, Kermeen (2007, 8–9) begins her book by writing of a trip to Haiti. She describes an excursion into the jungle and coming across a Voodoo ceremony. She writes in a way that echoes Robert Tallant's (1946) description of St. John's Eve in New Orleans. As Kermeen describes the incident,

> The voices grew louder and more intense as we came upon a clearing, where patches of sunlight fought through the thick canopy of vegetation to reveal about twenty-five scantily clad men stomping and dancing feverishly in a circle. They wore nothing but loincloths and heavy necklaces made of shells and bones, their bronze bodies glistening with sweat. Their faces were smeared with paint, making them look angry and scary, and they danced zealously round and round in a circle, some shaking primitive rattles made of unknown gris-gris, others waving long bones (animal bones, I hoped), chanting in a strange tongue. The longer it went on, the more wild and uninhibited they became.

Against her better judgment, she snaps a picture, and, as the camera flashes, the men lunge toward her before the guide leads her away. She barely escaped, she asserts, and "might now be carrying some diabolical curse" (13). As she opens the book with the ceremony, it frames the subsequent hauntings at the Myrtles Plantation (5; see, Miles, 2015, 99). In her tour, Miles (2015, 91), too, was told the story of Cleo, a Voodoo princess who lived on the nearby Solitude Plantation. Through ritual magic, she attempted to heal Ruffin Stirling's daughter who was suffering from yellow fever. His daughter did not survive the night, so Stirling and the plantation overseer "dragged the sleeping Cleo from the gilded room and into the front yard of the Myrtles, where they strung her up on an oak tree, asphyxiating her" (Miles, 2015, 91). Like with representations of Voodoo and Tatebe in Salem and Marie Laveau in New Orleans, through Cleo, Black spiritual beliefs continue to act as "exotic tidbits that enhance[] haunted story lines" (Miles, 2015, 119). Adding to the haunting qualities, Sillery (2001, 17) adds, "Local legend holds that the plantation, built on an ancient Tunica Indian burial mound, was cursed the moment the white man took over." Lyn Gibson (2015) informs of the connection between the Indian burial ground and subsequent hauntings:

> When construction began on the original home it was said that workers had unearthed a Native American burial ground. Bradford would order the

remains burned thus initiating over two centuries of hauntings that endure to this day. Bradford experienced great losses during his time at his family home. One of his sons would fall into the river while working one day, his body was never recovered.

The Myrtles, thus, exploits Voodoo and notions of indigenous spirituality for haunting effect (Miles, 2015, 124; Dickey, 2016).

Chloe's story also centers on Black violence. As with those enslaved in the attic of Madame LaLaurie, Chloe is an object of punishment, with representations of Black brutality and submission offered as thrilling spectacle (Miles, 2015, 91–93). When guides and television series speak of Woodruff cutting her ear off for eavesdropping, they participate in the spectacle of violence toward enslaved people. This is further symbolized through her turban and "stolen" earrings. Slave violence is evident, too, in descriptions of fellow enslaved people hanging her and throwing her in the river, as Sillery (2001), National Geographic (2008), and Pittman (2016) describe in lurid detail. In this regard, the Chloe postcard is reminiscent of a spectral lynching postcard (Apel, 2003). Given that photos of her are so easily displayed online, digital technology, as Miles (2015, 124) adds, "makes possible the easy duplication and dissemination of black women's suffering." Thus, much like the nameless masses in LaLaurie's attic or Julie freezing atop her home in New Orleans, Chloe's subjection is a key part of ghost tourism marketing.[5]

Finally, Chloe, like those enslaved, is an object, a commodity to be sold, "a thing and not a person" (Stevenson, 2021, 279). Her ghostly figure sells the postcard, and, while we did not see one in the gift shop, the plantation has sold a doll of her. Miles (2015, 118) describes the doll as haunted plantation kitsch, where "the bodies of black slaves" are "fodder for an innovating capitalist industry." To be sure, her image *also sells* the haunting experience, with her story told during each tour and her photograph being presented at the spot between the mansion and the General's Store where the mansion tours begin and end. She acts, in many ways, as the mascot for the Myrtles. Chloe is also put to work ideologically, as her story reinforces numerous controlling images of Black women as Jezebel, tragic mulatto, and mammy, and she is also described as childlike, conniving, and even uppity. To be sure, her ghostly figure is further put to work, as she has been described by tourists as having "a habit of floating about and tucking people in at night" (Miles, 2015, 98). She, thus, continues to serve slave duties into the afterlife (24).

Chloe is, thus, *captured* by the camera—or Photoshopped into marketable images—and is *put to work* for visitors, who hope to capture a glimpse of her. She is "the malleable black slave woman ghost," who "can appear as any visitor's fantasy" (Miles, 2015, 98). Such whitewashing continues with the second narrative that markets the Myrtles Plantation, the antebellum splendor narrative.

"Step into the Past for a Visit of Antebellum Splendor"

We wrote of our experience arriving at the Myrtles:

> We pull into the plantation in our rented car and are welcomed by crepe myrtles and oak trees. Once parked, we walk around the grounds. There is music playing and people, mostly white, relaxing on chairs. A Black woman is taking pictures with her electronic tablet, and a white man sits in leisure on the porch of one of the cottages. We walk to a gazebo in the middle of a pond, where weddings are hosted. An older white couple hold hands as they wander around the water's perimeter. Ducks play nearby, and a gecko climbs a tree. We sit on the porch of the mansion and await our first tour. It is serene but unsettling, as slavery haunts the landscape.

This note describes what we refer to as the antebellum splendor narrative, which glorifies Southern slave-owning society and has been famously depicted in the film *Gone with the Wind* (1939).[6] When relaxing on rocking chairs, or gazing at the pond, visitors are allowed to "be nostalgic for those bygone days of slavery" (Miles, 2015, 81–83; Figure 3.2).

The antebellum lifestyle is presented in promotional material. The Myrtles website invites prospective customers to "step into the past for a visit of

FIGURE 3.2 The Myrtles Mansion

antebellum splendor." It prompts the viewer to "relax in a giant rocker on the wide veranda or stroll through our historic grounds laced with Live Oak trees, Crepe Myrtle trees, azaleas and other flora and fauna typical of antebellum plantations." One can also dine on Southern cuisine at Restaurant 1796, ordering a Rebel Whiskey (named after original owner, David Bradford). The supplied map shows off the various establishments, such as the Caretakers Quarters, the Cottages, and the General's Store. The colorful map appears designed to look as though one is walking through a theme park.[7] Taken together, the website promises the visitor will bask in Southern antebellum splendor,

> The saga of the Antebellum South and a lifestyle that will never be for-gotten lives on at the grand mansion known as the Myrtles Plantation high on the hill in St. Francisville, Louisiana.

And it's not just the grounds—the mansion itself allows customers to indulge in antebellum architecture. The website informs prospective custo-mers that Ruffin Stirling, who purchased the property in 1834, "completed the mansion in the grandeur that one can see today." The veranda is "noted for its ornamental ironwork," and the stained-glass entrance is original to the house. The entrance foyer has "some of the finest examples of the faux-bois and open pierced friezework in existence today." In the center hangs a French chandelier made of "Baccarat crystal [that] weighs more than 300 pounds." For social life, Stirling constructed mirrored "ladies and gentle-men's parlors"; a dining room "to hold festive dinners and to discuss events of the day"; and a gaming room, which offers "a restful and intimate atmosphere for games of chance." Such rooms, the website assures, are "important to plantation life."

When touring the mansion, our guide described the architecture and social life, as she took us to the foyer, French room, dining room, ladies and gentle-men's parlors, and the gaming room. In the dining room, for instance, she visualized plantation life by showing us faux food, décor, and window drapes, and she spoke of a "servant boy" who would fan guests (Figure 3.3). We wrote,

> We are in the dining room, where the family ate the oleander cake. There is fake cake, sweets, and butter on the large wooden table. With humor, she assures our group that the cake is not real, a clear reference to Chloe's poi-soned cake. She talks about Stirling's wife, Mary Catherine Cobb, and how she displayed her wealth, although she does not mention where such wealth came from. She shows off the window drapes that touch the ground; it means, she explains, that they did not care about dirt, which was a sign of wealth. This is depicted in *Gone with the Wind*, she adds. She also points to a fly trap and salt which sit at each spot, "unless she didn't like you." She directs us to a corner of the room. Here, she adds, a "servant boy" would pull on a mechanism that rotated a ceiling fan to cool off guests. While not a ghost,

FIGURE 3.3 The Myrtles Dining Room

the imagined and nameless "servant boy," like Chloe, is put to work on the property.

Kermeen (2007, 30–31) describes such historical visualization in her own description of the dining room:

> In my head I flashed to an earlier time, when genteel ladies and gentlemen were seated at the table, toasting their host, each other, and the latest crop of cotton, delicately clinking the fragile glass.

To be sure, visitors can stay in one of the mansion's bedrooms, including plantation owner Clark Woodruff's room, which is the largest in the house. Visitors may also stay in the so-called Caretaker's Quarters, which our self-guided tour added was featured on the Travel Channel's *Ghost Hunters* (2005).

A customer review promotes the antebellum lifestyle marketed on the plantation. The writer recognizes the serenity afforded on the property, encouraging a trip, whether the plantation is "haunted or not."

> Haunted or not, this place deserves to be the center of any vacation tour of the South. We loved just sitting and enjoying a quiet evening experiencing the Myrtles Plantation. It was the most friendly staff I can say I have seen in a long time. Everyone went out of their way to make sure we were comfortable and welcome.

Ultimately, the antebellum splendor narrative reinforces bucolic imagery of plantation life, while slavery is largely represented through Chloe's ghost story which is rife with racist stereotypes. Through these dual narratives, the tourist embraces "white nostalgia" as "a mode of remembrance celebrating a specific time and place in history by erasing narratives of racism and by whitewashing memories" (Adamkiewicz, 2016, 17; Morris & Arford, 2019, 440). Yet, even in the ostensibly serene landscape, a deeply unsettling feeling remains—that of slavery haunting the property. In the final section, we read Chloe's story against the grain of these narratives to critically engage with the structural violence committed on the plantation and how such unresolved violence permeates today (Hartman, 2022).

"A Slave Girl Standing Between"

In what ways does Chloe's story tell us about slavery, rather than obscure it? Importantly, the fact that an enslaved person is discussed at all, even as a whitewashed figure, shows that, *slavery did exist on the plantation*. While our guide did not mention slavery, for instance, when discussing that Woodruff transitioned from indigo to cotton production, such production must come from someone. We were, after all, on plantation row, where mechanical technology drove the importation of enslaved people from the Upper South (Miles, 2015, 103). In 1820, David Bradford's wife, Elizabeth, owned 24 enslaved people and Clark Woodruff owned 5. In 1830, Elizabeth Bradford owned 10 and Woodruff owned 33. By 1830, it is believed that Woodruff may have owned 480 enslaved people. The Stirlings together owned 173 enslaved people (see Miles, 2015, 87; Pittman, 2016).[8] The banality of the slave trade appears in a letter David Bradford wrote to his friend David Redick of his wife, Elizabeth, purchasing enslaved people when he was in Pennsylvania for business. He penned,

> On my arrival I found my family well & my plantation affairs better conducted than if I had been at home. Mrs. B has acquired high reputation as a cotton planter. She bot. an excellent negro wench in my absence – Sent money Towland & had four more which arrived here after my arrival. I bought two as I passed Wheeling – 1 of Col. Chaplain a young wench – a 2nd of Mr. South a lad of 16 years of age
>
> *(Pittman, 2016, 44).*

Of note, this letter is excerpted in Rebecca Pittman's (2016) *The History & Haunting of the Myrtles Plantation*, which we purchased in the Myrtles General's Store. While Pittman does not generally condemn the Myrtles or such tourism, she does offer historical documentation, opening an avenue for critical inquiry on the all-too-ordinary horrors of slavery, as it existed right on the same grounds where one can now purchase a "Rebel Whiskey" in Bradford's honor.

Chloe also represents violence imposed on Black enslaved women. In rejecting the Jezebel trope, we can see more clearly how her story reflects planters' sexual abuse of enslaved women to maintain power. The view of Black female hypersexuality made Black women, whether enslaved or not, as considered virtually unrapeable (Crenshaw, 1991; Roberts, 1998).[9] As Patricia Hill Collins (2008, 89) elaborates,

> Jezebel's function was to relegate all Black woman to the category of sexually aggressive women, thus providing a powerful rationale for the widespread sexual assaults by White men typically reported by Black slave women.

Between 1728 and 1776, none of the nearly 100 reports of rape or attempted rape in 21 newspapers in nine American colonies reported the rape of a Black woman (Block 2002; Kendi, 2016, 209).[10] Importantly, state-sanctioned sexual violence toward Black women and gender-nonconforming people has continued in prisons and jails through institutionalized strip searches (Davis, 2003, 80).

In this regard, Black women are viewed as not having a "self" worth defending (Kaba, 2023, 10–11).[11] Mariame Kaba (2019) draws a connection between the experiences of two Black women, Celia and Marissa Alexander. Celia was a 19-year-old enslaved woman in Missouri. After enduring five years of sexual violence at the hands of her master, she killed him on June 23, 1855. Yet, the court concluded that, since Celia was property, she "was not considered a person under the law and could therefore not be raped." She was found guilty of murder and sentenced to death by hanging. In 2010, Alexander fired a single warning shot from her licensed gun when confronted by her abusive, estranged husband. She had just given birth to a daughter nine days earlier. Marissa was sentenced for 20 years and, only after a nationwide legal defense campaign, was set free after five years of incarceration and two years of house arrest. Chloe's ghost stands with Celia and Alexander, as a Black woman who was killed for committing an act of survival. And, when Black women resist, such as Chloe did, they are viewed as assailants—as conniving, thieving, and dangerous.

Like with Tatebe, we can also locate resistance through the figure of Chloe. While described as having a "nasty habit of eavesdropping," Chloe's use of information, secrets, and spying speaks to a broader strategy of survival. Patrisse Cullors (2021, 24), who is the cofounder of Crenshaw Dairy Mart, Black Lives Matter, and Dignity and Power Now, reflects on her own keeping of secrets: "As a young, poor, Black girl, I learned that keeping secrets was largely about my survival and safety." Theft, such as that of the earrings, can be considered an act of defiance and survival for enslaved people. Even if she had not intended to kill Woodruff's children, the vision of the rebellious slave also haunted planters' minds—from Haiti to the East Bank of the Mississippi River where the German Coast Uprising occurred (Mbembe, 2019, 166). Indeed, through Chloe, the antebellum South can be reimagined not as a place of white splendor but one of constant and collective resistance to the slave system (Page & Woodland, 2023).

Like in Salem and New Orleans, guides may offer critical engagement. Miles (2015, 109–115) tells of Tommy, a young African American guide who was a descendant of enslaved people who worked on the plantation. While his grandmother was a Voodoo priestess, he did not go into detail when asked about Voodoo by a member of the group. He did not speak negatively of Chloe, merely reporting, "You'll see her shadow or feel a chill on your arm or shoulder. . . . Just let her pass. She'll do no harm." He noted with humor that he played her on *Discovery Channel* and *Travel Channel*, "Look at the face, y'all," he said, sweeping his long hair back. "Look at the face." The audience also interacted positively with him when he criticized Woodruff's sexual abuse,

> The judge was not there to eat the cake because he was, Tommy said with a knowing inflection, "taking care of some business out in the cotton fields." In response to his sexual innuendo, the other black woman in the group responded "Mmm hmm," a sarcastic cultural vernacular rife with moral judgment. A white woman chimed in: "Chloe should have slapped the hell out of that judge."

Miles concludes that Tommy displayed a transgressive portrayal, enacting a "queer vernacular performance" (Johnson, 2001a), and he used "the staircase of an old plantation house as his stage." He "initiated a call-and-response exchange," encouraged "outspoken feedback on Chloe's history," and began the tour by reminding audiences not to believe everything they hear. There are still issues with his performance, primarily that he is a Black man speaking of violence against Black women in the 19th century with "a cheekiness that conveyed humor" within an overall exploitative industry. For some visitors, we speculate, Tommy may have also been rendered through the "fun and frolic" trope (Hartman, 2022, 7). A white woman told Miles, "Everybody loves Tommy. People request him. He's a free spirit. You can't tie him down." On one hand, his not being "tied down" renders him a free spirit—the antitheses of an enslaved person. On the other hand, he is committed to a kind of light-hearted engagement palatable for white visitors. Nevertheless, his ability to challenge dominant scripts, like that of our Black tour guide in New Orleans, affords an oppositional reading necessary for a critical ghost tourism.

Such critical interpretations of the mansion can also be found online. We began *America's Horror Stories* with the photo uploaded on December 4, 2017 to the Myrtles Facebook of several white women and a ghostly figure of an enslaved person in the background. "Not your average guest selfie," the post's caption read. While it did receive positive media attention and numerous "likes" and "shares," not all comments were supportive. From a brief overview of the comment section, we identified several remarks on the photo's exploitation. A reply post read, "Continued profit from other people's pain. The lack of empathy is astounding!" Another added, "Spirits of the ancestors who died while enslaved on that plantation." One described the photograph as exploiting "a pained history."

> It may have been 300 years ago, but you people have never shown slavery the due remorse it should have been shown.
> There was no 40 acres and a mule . . . racism and modern slavery still exist, and y'all are taking selfies with photoshopped slave kids in 2017.
> And we were not sold into slavery by our own, we were coerced and threatened and bribed by "your own.". Either way doesn't make it right and doesn't mean you get to take selfies of such a pained history without our opinion.

Chloe therefore stands not just "between two of the buildings on the plantation," as promotional material has it (Figure 3.4), but between past slavery and present

FIGURE 3.4 Space between Buildings Where Chloe Was Photographed

racist violence—of a social violence left unresolved (Gordon, 2011, 2). Indeed, Louisiana has the highest incarceration rate in the United States at 1,067 per 100,000 residents, which is more "than any democratic country on earth" (PPI). It also hosts the plantation prison of Angola (see Kennedy, 2017). Thus, one can see in Chloe's story, as with Toni Morrison's (1987) characterization of Beloved, a symbolic bridge "between that of the living and that of the dead." The symbolic bridge, as Mohammad Deyab (2016, 16) adds, "connects past with present, reality with fantasy, life with death." As such, Chloe also represents resistance in her actions and of those enslaved more generally, as well as by tour guides who disrupt the dominant whitewashed framing of the plantation.

Conclusion—Escape from the Myrtles Plantation

The Myrtles Plantation offers two whitewashed narratives: of the haunting narrative and the antebellum splendor narrative. The haunting qualities of the landscape and past violence are in turn plotted into a manageable, marketable—and thrilling—ghost experience, while offering an image of a romanticized antebellum South. The home page of the website reads, "Escape to the Myrtles Plantation," obscuring the reality that the grounds were a place where so many would desire to *escape from*. The ghost tour narratives, thus, erase, entice, and offer a representation of Chloe that reinforces and animates state violence today.

In the final chapter, we make our way back north, to Gettysburg, Pennsylvania, where a key battle of the Civil War was fought, a war that resulted in the abolition of the slave system, which had brought so much material wealth to Louisiana and to the United States as a whole. And, as we drive off the premises, we make note of a store across the Blues Highway. It is named the Plantation Feed & Supply.

Notes

1 We took screenshots for archival purposes (screenshots on July 7, 2022) from the website, www.myrtlesplantation.com/. Overall, the site, as a digital vista for prospective customers, acts as visual/textual rhetoric. (Daniels & Gregory, 2016). It should be noted that, within the last two years, the Myrtles overhauled their website, so our analysis is of the previous site—perhaps representing a haunting digital archive.
2 A 1941 Louisiana state tourism guide does inform that "several ghost stories are centered on the Myrtles" (Miles, 2015, 104).
3 Perhaps the most representative of this is that of Saartjie "Sara" Baartman, an African woman who was displayed in Europe as the "most correct and perfect Specimen of that race of people," that of the so-called Hottentot. By contrast, Anglo-Saxon women were deemed as the ideal beauty, based on English svelteness and Anglo-American slenderness (Strings, 2019, 91–98).
4 As a note, as Chloe is also rejected by Woodruff, her story incorporates the tragic mulatto trope (Miles, 2015). Like Julie in New Orleans, she is rejected by the white man of her desire and is unable to fit Southern ideals of white womanhood (Foley, 2021, 16).

5 To add, a guide told Miles (2015, 92) that a film producer who stayed in the French Bedroom saw Cleo hanging from the ceiling, "as vividly as 'a real lynching.'"
6 Indeed, Kermeen (2007, 14) speaks of being mesmerized by *Gone with the Wind*, which had inspired her to run a Southern inn.
7 Ken Anderson considered Louisiana plantations when designing Disney's Haunted Mansion ride, which is located in the theme park's New Orleans Square (Coles).
8 These are according to census records, although the "480" number is from Pittman (2016), who does not include detailed footnotes, as explained by Miles (2015, 7).
9 Slave codes also designated slave status as passed down maternally. The 1662 Act XII of the Virginia House of Burgesses states "that all children borne in this country shall be held bond or free only according to the condition of the mother" (Morgan, 2022). This further encouraged the raping of enslaved women.
10 Indeed, not only did slave codes permit sexual assault, but they tightly *regulated* interracial relationships. In 1630, the Virginia colonial court ordered the whipping of white man Hugh Davis for "defiling his body in lying with a negro" (Oluo, 2022, 11). In 1640, a white man was brought to law for impregnating a Black woman: it was the woman who was whipped this time, and the man was sentenced to church service (13).
11 As Mariame Kaba (2023, 10–11) writes, "If black women's bodies can always be violated and if black women are easily killable, then the notion of self-defense can never apply. Black women do not have a 'self' worth defending."

To be sure, this links to a longstanding judicial and extrajudicial regulation of Black self-defense. For instance, the *Dred Scott* (1857) decision warned that recognizing Black people as citizens would allow them "to keep and carry arms wherever they want" (Jackson, 2022, 56). Henry Adams (1880), a Louisiana freedman, veteran, Republican canvasser, and community organizer, also stated of white retribution toward Black self defense, as documented by Kadida Williams (2023, 226):

> If one Black person killed a white person, even in self-defense, whites mobilized to "kill fifty colored men for the one white one," Adams said. Then, whites would go down "by the fifties and hundreds" to any weapons store to see whether Black people were trying to acquire arms. And when white people began rampaging through communities and Black neighborhoods, "Colored men [could not] buy ammunition" to properly defend themselves.

References

Adamkiewicz, Ewa A. 2016. "White Nostalgia: The Absence of Slavery and the Commodification of White Plantation Nostalgia." *Aspeers: Emerging Voices in American Studies*, 9 (1):13–31.

"America's Spookiest Homes: Myrtles Plantation." 2014. *HGTV*.

Apel, Dora. 2003. "On Looking: Lynching Photographs and Legacies of Lynching after 9/11." *American Quarterly*, 55(3): 457–478.

Barbot, Jean. 1732. *A Description of the Coasts of North and South-Guinea*. London.

Bernier, François. 1684. "A New Division of Earth." In *History Workshop Journal*, edited by Siep Stuurman, 50.

Block, Sharon. 2002. "Rape and Race in Colonial Newspapers, 1728–1776." *Journalism History*, 27(4): 146–155.

Coles, Sasha. "The Haunted Mansion: Gore, Ghosts, and the Greek Revival." *The Enchanted Archives*. https://enchantedarchives.com.

Collins, Patricia Hill. 2008 [1990]. *Black Feminist Thought: Knowledge, Consciousness, and the Politics of Empowerment*. New York and London: Routledge.

Crenshaw, Kimberlé. 1991. "Mapping the Margins: Intersectionality, Identity Politics, and Violence against Women of Color." *Stanford Law Review*, 43(6): 1241–1299.

Cullors, Patrisse 2022. *An Abolitionist's Handbook: 12 Steps to Changing Yourself and the World*. New York: St. Martin's Press.

Daniels, Jessie, and Karen Gregory. 2016. *Digital Sociologies*. Bristol: Policy Press.

Davis, Angela Y. 2003. *Are Prisons Obsolete?* New York: Seven Stories Press.

Dickey, Colin. 2016. *Ghostland: An American History in Haunted Places*. Westminster: Penguin.

Deyab, Mohammad ShaabanAhmad. 2016. "Cultural Hauntings in Toni Morrison's *Beloved* (1987)." *English Language, Literature & Culture*, 1(3): 13–20.

Duncan, Cynthia. 2016. "Creating and Maintaining the Plantation World in the Mississippi Delta." *In These Times*. https://inthesetimes.com.

Feimster, Crystal N. 2022. "Lynching." In *Four Hundred Souls: A Community History of African America, 1619–2019*, edited by Ibram X. Kendi and Keisha N. Blain, 254–257. New York: One World.

Ghost Hunters. 2005. "Myrtles Plantation." *Travel Channel*.

Gibson, Lyn. 2015. Haunted Louisiana—the Myrtles Plantation.

Gone with the Wind (1939).

Gordon, Avery. 2011. "Some Thoughts on Haunting and Futurity." *Borderlands*, 10(2): 1–21.

Hartman, Saidiya. 2022. *Scenes of Subjection: Terror, Slavery, and Self-Making in Nineteenth-Century America*. New York: W.W. Norton & Company.

Jackson, Kellie Carter. 2022. "The Virginia Law That Forbade Bearing Arms; or the Virginia Law That Forbade Armed Self-Defense." In *Four Hundred Souls: A Community History of African America, 1619–2019*, edited by Ibram X. Kendi and Keisha N. Blain, 55–56. New York: One World.

McElya, Micki. 2007. *Clinging to Mammy: The Faithful Slave in Twentieth-Century America*. Cambridge: Harvard University Press.

Morgan, Jennifer L. 2022. "Elizabeth Keye." In*Four Hundred Souls: A Community History of African America, 1619–2019*, edited by Ibram X. Kendi and Keisha N. Blain, 39–42. New York: One World.

Morrison, Toni. 1987. *Beloved*. New York: Knopf.

Johnson, Patrick E. 2001a. "'Quare' Studies, or (Almost) Everything I Know about Queer Studies I Learned from My Grandmother." *Text and Performance Quarterly*, 21(1): 1–25.

Johnson, Walter. 2001b. *Soul by Soul: Life inside the Antebellum Slave Market*. Cambridge and London: Harvard University Press.

Kaba, Mariame. 2019. "Black Women Punished for Self-Defense Must Be Freed from Their Cages." *The Guardian*. https://www.theguardian.com.

Kaba, Mariame. 2023. "Introduction: We Can Only Survive Together." In *Let This Radicalize You*, edited by Kelly Hayes and Mariame Kaba, 8–15. Chicago: Haymarket Books.

Kendi, Ibram X. 2016. *Stamped from the Beginning: The Definitive History of Racist Ideas in America*. Boston: Bold Type Books.

Kennedy, Liam. 2017. "'Today They Kill with the Chair Instead of the Tree': Forgetting and Remembering Slavery at a Plantation Prison." *Theoretical Criminology*, 21(2): 133–150.

Kermeen, Frances. 2007. *The Myrtles Plantation: The True Story of America's Most Haunted House*. New York: Warner Books.

Mbembe, Achille. 2019. *Necropolitics*. Durham: Duke University Press.

Miles, Tiya. 2015. *Tales from the Haunted South: Dark Tourism and Memories of Slavery from the Civil War Era*. Chapel Hill: University of North Carolina Press.

Morris, Patricia, and Tammi Arford. 2019. "'Sweat a Little Water, Sweat a Little Blood': A Spectacle of Convict Labor at an American Amusement Park." *Crime, Media, Culture*, 15(3): 423–446.

National Geographic. 2008. *National Geographic's Myrtles House: The South's Spookiest House*. https://www.nationalgeographic.com.

Morrison, Toni. 1987 [2019]. *Beloved*. New York: Vintage Books.

Moses, Jennifer. 1999, July 25. "More Than Just a Place to Stay: Louisiana; Ghosts and Atmosphere Galore in a Restored Plantation House." *New York Times*.

Newman, Rich. 2021. "The Myrtles Plantation. The History and How to Visit." *Passport to the Paranormal*. https://www.llewellyn.com.

Oluo, Ijeoma 2022. "Whipped for Lying with a Black Woman." In *Four Hundred Souls: A Community History of African America, 1619–2019*, edited by Ibram X. Kendi and Keisha N. Blain, 11–14. New York: One World

Page, Cara, and Erica Woodland. 2023. *Healing Justice Lineages: Dreaming at the Crossroads of Liberation, Collective Care, and Safety*. Berkeley: North Atlantic Books.

Petersen, Amanda M. 2024. "Community-Oriented Copaganda: Anti-Black Violence in a Visual Archive of Policing." *Crime, Media, Culture*. doi:17416590241231904..

Pittman, Rebecca F. 2016. *The History and Haunting of the Myrtles Plantation, 2nd Edition*. United States of America: Wonderland Productions.

Potter, Amy E. 2016. "'She Goes into Character as the Lady of the House': Tour Guides, Performance, and the Southern Plantation." *Journal of Heritage Tourism*, 11(3): 250–261.

Ralph, Laurence. 2022. "The Code Noir." In *Four Hundred Souls: A Community History of African America, 1619–2019*, edited by Ibram X. Kendi and Keisha N. Blain, 57–61. New York: One World1.

Roberts, Dorothy. 1998. *Killing the Black Body: Race, Reproduction, and the Meaning of Liberty*. New York: Vintage.

Sillery, Barbara. 2001. *The Haunting of Louisiana*. New Orleans: Pelican Publishing.

Stevenson, Bryan. 2021. "Punishment." In *The 1619 Project: A New Origin Story*, edited by Nikole Hannah-Jones, 275–283. New York: One World.

Strings, Sabrina. 2019. *Fearing the Black Body: The Racial Origins of Fat Phobia*. New York: New York University Press.

Tallant, Robert. 1984. *The Voodoo Queen*. New Orleans: Pelican Publishing.

Unsolved Mysteries. 2001. "Myrtles Plantation."

Vaughn, Holley Ann. 2012. *A Critical Ethnography of The Myrtles Plantation in St. Francisville, Louisiana with Ruminations on Hauntology*. Baton Rouge: Louisiana State University.

Wells, Ida B. 1982. *Southern Horrors: Lynch Law in All Its Phases*. Boston: Bedford Books, 1997.

4

CONJURING THE CONFEDERACY

Introduction

The Civil War (1861–1865) is often considered by historians as the bloodiest conflict in American history, as over 620,000 died over the course of the conflict, With the election of Abraham Lincoln in 1860, Southern states began to secede out of fear that the federal government would abolish slavery. South Carolina was the first to do so on December 20, 1860. Military conflict arose in April 1861, when Southern forces opened fire on the federal base, Fort Sumter, in Charleston Harbor, South Carolina. Within two years, the North and South engaged in battles such as Manassas (1861), Antietam (1862), Fredericksburg (1862), and Chancellorsville (1863). The 1863 Battle of Gettysburg was a result of Virginia General Robert E. Lee's second attempt to invade the North after a draw at Antietam in 1862. The Northern invasion, he surmised, would offer his army supplies and put soldiers en route to Philadelphia (Guelzo, 2013, 19).

Opening shots were fired on July 1 on Seminary Ridge, located north of the town. The Confederates won the battle. On July 2, armies converged on the farmlands of Gettysburg and fought over control of the high ground of Little Round Top and Devil's Den, with the Union Army declaring victory. On July 3, the battle ended with the ill-fated Pickett's Charge on Cemetery Hill. The Union was positioned atop the hill. Confederate soldiers marched in the open with no protection from opposing gunfire. Pickett's Charge resulted in Union victory, putting an end to Lee's northern invasion (see Catton, 1963). Over the course of the three-day battle, more than 50,000 soldiers either died, were wounded, or went missing. One-third lost their lives, and over 1,000 bodies remain unaccounted for and are possibly still buried on the battlefield—which spans the town and outer limits. The Battle of Gettysburg was, as Allen Guelzo (2013, 5) describes, "[T]he greatest and most violent collision the North American continent had ever seen."

DOI: 10.4324/9781003397809-5

The scope of atrocity in the Civil War led to a cultural fascination with death. Indeed, the death toll inspired a rise in the spiritualist movement, where many hoped to contact their fallen loved ones (Manseau, 2017). The popular book *Gates Ajar* (Phelps, 1868), too, conceived of heaven as "a more perfect Earth," thus offering the sense that fallen soldiers were merely waiting to meet again in a world much like our own (Faust, 2008, 187). Mid-19th-century developments in photography also brought the dead to the public eye, as battle photographers like Alexander Gardner published photos of fields littered with bodies (Manseau, 2017, 122). As Tony Horwitz (1998, 386) reminds us, "Without photographs, rebs and Yanks would seem as remote to modern Americans as Minutemen and Hessians."

Certainly, the fascination with the Civil War, and with Gettysburg in particular, remains. Gettysburg has long been the country's most visited battlefield site, attracting over two million tourists a year (Conservation Fund, n.d.). The Gettysburg ghost tourism industry has subsequently catered to visitors' interests in war, death, and the afterlife. Local ghost historian and park ranger Mark Nesbitt has declared, "Gettysburg may very well be, acre for acre, the most haunted place in America!" (Mark Nesbitt's Ghosts of Gettysburg, 2022). Tour companies offer walking, bus, and self-guided tours and sell a variety of Gettysburg ghost books, including Nesbitt's eight-part series, *Ghosts of Gettysburg* (1991–2018); Joe Svehela's three-part series, *Ghostly Images* (2004–2007); and Johlene "Spooky" Riley's *Ghostly Encounters of Gettysburg* (2011) and *Ghost Hunting: The Gettysburg Files* (2014). As a tour guide told our group, "You just can't get away from ghosts here in Gettysburg."

What do Gettysburg ghost stories say about racism, colonialism, and slavery? After all, as we have considered throughout this book, "remembering the dead is a matter of deliberation and craft" (Nudelman, 2015, 3). When guides speak of dead Civil War soldiers, they do more than merely provide historical description—they create our sense of social reality, historically and today. In this chapter, we identify three narratives in the industry that, we argue, downplay the role of slavery in the conflict and offer a sympathetic portrayal of the Confederacy. First is the atrocity narrative, where guides tell of blown-off limbs and corpses strewn about the battlefield. In doing so, they depoliticize the conflict by highlighting generalized human tragedy. Second is the reconciliation narrative, which displays the North and the South as having now found national forgiveness and unity. This diminishes slavery's role in the war and presents both sides as fighting for a noble cause. The third is the lost cause narrative, which glorifies the Southern fight and recognizes the heroism of Confederate soldiers (see Gallagher, 2008). Through these dominant narratives, tourists can engage with historical atrocity without having to meaningfully consider the American horrors of slavery, colonialism, and white supremacy.

Atrocity Narrative

The following note is from a self-guided audio ghost tour atop Cemetery Ridge, located on the upper reaches of the battlefield:

> We enjoy the summer breeze in our rental top-down jeep, while visitors traverse the space. As the narrator explains through the jeep's speakers, the last day of the battle took place in this location when the Confederates initiated Pickett's Charge. The Union held territorial advantage, as Confederates approached from the open field below. The narrator Mark Nesbitt (see 2022, 4) marks the atrocity of the Civil War in the recording,
>
> "Men were shot in the faces and brains and genitals and guts. They were stabbed by the particularly savage triangular bayonet, through the eyes and spine and throat."
>
> He then tells a ghost story, of a colonel and father-in-law who heard strange military noises from the field,
>
> "Desperate orders shouted . . . steel rammers ringing in muskets . . . the click of hammers cocked . . . the hoarse trill of a bugle . . . the clanking of artillery chains . . . a roar . . . shrieks . . . men gagging, crying, screaming, moaning, moaning, moaning" (Nesbitt, 2022, 49).
>
> The lush sun swept grounds now appear filled with phantom bodies of the dead.

The audio tour here highlights the atrocity of the battle, of the ghostly noises of "men gagging, crying, screaming."

While Gettysburg, and the Civil War more generally, was certainly filled with death, such atrocity must nevertheless be framed: it must be accounted for through a cultural register. The photographs of the battlefield dead taken by Alexander Gardner, for instance, represented "a nation divided" (Manseau, 2017, 135–137).[1] In Abraham Lincoln's 1863 Gettysburg Address, the president framed the death toll as defining a "new birth of freedom" and asserted that "these dead shall not have died in vain." Two years later, in his 1865 second inaugural speech, Lincoln questioned if the scope of death was divine punishment for the sin of slavery. (An elderly enslaved woman put it sharply, "It was a-comin. I allers 'spected to see white folks heaped up dead" [Faust, 2008, 54]).

In this section, we explore how guides utilize the atrocity narrative when describing death and disease on the battlefield, in field hospitals, and from burial. To reiterate, the atrocity narrative, as told in the ghost tourism industry, evinces a general claim of human suffering and potential toward violence, to which all soldiers were susceptible. Stories of amputated limbs and bloated corpses present an ultimately depoliticized, abstract soldier. Whether Union or Confederate, they are garnered similar sympathy as a totality, flattening political meaning behind what soldiers represent—national identity, statehood, and the struggle over slavery and white supremacy. Such depoliticizing, too, allows consumers to engage with the thrilling violence of the Civil War, while simultaneously disconnecting from the battle's links to slavery and colonialism, in turn presenting a (white) nation *unified* underneath the banner of past atrocity.

Not just in stories of battlefield conflict, the scope of atrocity is told in ghost stories of field hospitals. Field hospitals are establishments, such as schools, churches, and farmhouses, where injured and maimed soldiers were brought for medical attention. Standing in front of a school-turned-field-hospital for Union soldiers, our guide painted a vivid scene. Without modern medicine, he explained, doctors were required to amputate limbs, and they threw arms and legs into piles outside the windows. "Dried blood filled floors," he added, which, "You can still see with a black light." One tourist found in their pictures, he added excitedly, a surgeon with a hacksaw peering out one of the windows. He shared a rather convincing image to our group and reminded us to take pictures and check them out when we return to our hotels. In a Confederate field hospital next door, at an Old German Church, a guide also informed us that blood would seep through the floorboards, soaking those underneath it. In a haunted objects museum, a guide directed us to *feel* field hospital atrocity by touching a plank of wood from an operating table. The blood had ionized, giving it a brownish coloring, she explained. We wrote,

> Participants are invited to touch a piece of wood (blood stained) from numerous Civil War amputations to get "feelings." The wood has marks where soldiers dug their nails into the piece during the operation as their limbs were being sawed off by the surgeon. Our guide boasts about the sensation of being in the space of a Civil War hospital and discussed how past museum visitors experienced "flashbacks" of being in that space. They heard moaning, and feeling coldness on parts of their body where, presumably, the operation took place. In this, we are invited in the space in 1863, with its tension, dual ideologies, and white supremacy, to reclaim what was lost.

As with the battlefield and field hospitals, the scope of atrocity is also portrayed when guides describe the conditions of burial. In the previous example, where we stood in front of the school-turned-field-hospital, the same guide spoke of how, after the battle, the wounded and dead littered the fields and town. As the Union was unable to bury the deceased quickly, bodies began to bloat and burst in the hot July sun. Town residents even applied rose nip or peppermint oil under their noses to mask the smell of death, and starving hogs resorted to feeding on bodies, with some soldiers perhaps conscious at the time. Soldiers, he continued, were buried in shallow graves, and, once rain fell, the bodies resurfaced. When at Plum Run Creek, another guide spoke of a flood where bodies, arms, and legs flowed through the stream. In the area, he added, a Confederate reenactor, who was honoring his ancestor, witnessed an apparition of a Confederate soldier who had blood on the sleeve of his coat. Visitors, he noted, have similarly seen orbs and shadows near the creek.

As we have stated, the Civil War certainly was violent.[2] Our guides did describe conditions documented by historians. Bruce Catton (2013, 84) lists several accounts by soldiers:

> Confederates disappeared in a boiling cloud of dust and smoke, in which knapsacks and muskets and horrible fragments of human bodies were tossed high in the air; one Federal soldier remembered that there came from this part of the field a strange sound that was like an agonized gasp of pain coming from hundreds of throats. No one seems to have remembered hearing any cheers from either side. One soldier recalled only "a vast mournful roar" that seemed to rise from the entire field.[3]

Catton (1963, 133) adds of Gettysburg,

> Indeed, the town was hardly fit to live in, because the air was heavily tainted with foul odors. Thousands of dead bodies had lain under the July sun for days, and although burial squads were kept busy it took a long time to get all of the pitiful remains under the sod.

Union General Ulysses S. Grant also wrote of the scope of death after the Battle of Shiloh in Tennessee,

> I saw an open field . . . so covered with dead that it would have been possible to walk across the clearing, in any direction, stepping only on dead bodies without a foot touching the ground
>
> *(Faust, 2008, 58).*

Once in the field hospitals, soldiers were indeed subject to amputation and disease, as medical providers did not wash their hands to preserve water and because germ theory had not yet developed (4). As Drew Gilpin Faust (133) adds, "Overworked surgeons became numb with sheer fatigue as they worked all day long, day after day, amputating arms and legs." Witnesses at field hospitals

> almost invariably commented with horror on the piles of limbs lying near the surgeon's table, disassociated from the bodies to which they had belonged, transformed into objects of revulsion instead of essential parts of people
>
> *(xvi).*[4]

While guides speak of such atrocity within the ghost tourism industry, they must also frame the mass death in a way that appeals to ghost tour consumers. As we argue, the atrocity narrative tells of soldiers, whether Union or Confederate, as a depoliticized totality—configured through descriptions of bodies, limbs, and bloodshed. In doing so, guides portray Civil War atrocity without offering controversial political inquiry. The question of slavery and nationhood remains a background assumption. Consequently, they implicitly feed into, or at least do not challenge, pro-Confederate sympathy, which we will explore more deeply in the next two sections. To this, we move to a similarly depoliticizing framing of the Civil War, the reconciliation narrative.

Reconciliation Narrative

To celebrate the 50th anniversary of the battle in 1913, two veterans, one Union and one Confederate, shook hands across the stone wall on Cemetery Ridge. In 1938, President Franklin D. Roosevelt opened the Eternal Light Peace Memorial. Two Civil War veterans, also one Union and one Confederate, assisted in the unveiling of the monument. Depictions of such national unification persist in the burial of soldiers. The 1867 National Cemetery Act mandated that the Secretary of War provide "proper burial" for those who fought in the war. While separated in the cemetery, the collective burial effort served to unite the nation through the figure of the soldier, whether Union or Confederate. Walt Whitman penned in 1882, "the dead, the dead, the dead—*our* dead—or South or North, ours all (all, all, all, finally dear to me)" (Faust, 2008, 269). Through such commemoration and burial efforts, one can see the federal government's attempt to present national unity in the aftermath of the Civil War.

The reconciliation narrative emphasizes unity between Union and Confederate soldiers. It celebrates the ideals and missions of both parties and, thus, promotes Confederate valor. Like the atrocity narrative, it also does not mention slavery (Gallagher, 2008, 33). Indeed, the 2024 161st anniversary battle reenactment is scheduled to have talks by reenactors posing as Confederate Generals Jubal Early and Bedford Forrest (who was the first leader of the KKK and was not present at the Battle of Gettysburg). In this section, we explore two aspects of the reconciliation narrative in the Gettysburg ghost tourism industry. First, we consider a ghost story where Union and Confederate friends were *unable to reunite* after battle, thus offering a tragic ending that reinforces the importance of North–South unity. Second, we consider how unity between the Union and Confederacy presents a sense of friendly competition, which is made marketable to sell tourist merchandise, including ghost memorabilia.

A commonly told story, both in ghost books and by guides, is that of three fallen friends: Union soldier Jack Skelley, Confederate soldier Wesley Culp, and Gettysburg resident Jennie Wade. Numerous guides offered the basic story to us: Skelley, Culp, and Wade were childhood friends. Skelley and Wade were secretly engaged. Culp was an apprentice in the town's carriage maker shop and, when the owner relocated to Virginia, he followed. At the outbreak of war, Culp joined the Confederate army, and Skelley enlisted with the Union. Skelley was severely wounded at the Second Battle of Winchester in 1863. Culp came across him at a field hospital. Skelley asked him a favor: that he deliver a letter to his beloved, Jennie Wade. Culp obliged; however, on July 2, he was mortally wounded at Gettysburg on his uncle's farm, Culp's Hill. The next day, at the battle's end, Wade was also killed by Confederate fire when baking bread for Union soldiers at her sister's home. The bullet went through two doors and hit her in the back, piercing her heart before affixing in her corset. Skelley's message was never delivered, and he died on July 12.

Guides described the three friends in the afterlife. One stated that Skelley and Wade now reside together in Wade's house, as some visitors have felt their presence on a spirit ghost box. (He also pointed to a small window on the building, stating it was perhaps a spot on the underground railroad, which was a fleeting, yet welcome reference to slavery.) He continued that several also have seen a man in a Confederate outfit outside of Jennie Wade's sister's house, where Jennie was shot. The soldier, who is presumably Culp, will approach onlookers, dig through his pocket, and, before presenting a letter, disappear. Or, perhaps, as Mark Nesbitt (2022, 8) ponders, "somewhere, in the weird world beyond, three friends still seek forever one another and an answer to the unanswerable." The drama focuses on friendship, love, and tragedy, thus blurring political boundaries between the North and the South. Indeed, this sense of tragedy is also depicted in a memorial of Wesley and his brother William, a Union soldier. William survived the war, and it is unknown if they met each other during the conflict. Nevertheless, the Culp Brothers Monument honors the torn-apart brothers' legacy. Within the human bonds described in these stories, of friendship and brotherhood, the politics surrounding the war are lost—and, in the afterlife of war, such reconciliation is deemed aspirational.

Through eventual unification, the reconciliation narrative also permits an amalgamation of pro-Confederate and pro-Union marketing through a conception of friendly competition.[5] One may eat at Confederate Pickett's Buffet for lunch; stop by the Lincoln-named Fourscore Beer Co. for a beer; grab a Yankee or Rebel Cola from a tourist shop; dress up as a soldier of either side for an "Old Tyme" photo; or go to the Blue & Gray restaurant for dinner, where they may purchase a Confederate Burger, a Union sandwich, or even some Ghost Wings. Guides dress as Union and Confederate soldiers, and some tourists do the same. The significance of choosing sides was apparent at the end of one tour. Our Confederate-dressed guide, who regularly spoke of his reenactment work, offered gifts to the children at the end of our tour. For one child, he asked humorously if they were more like a "Yank" or a "Rebel." The child looked sheepish and a bit confused, and his parents looked similarly. After a pause, the guide offered the child a Confederate belt buckle, to which he held excitedly.

In cultivating a sense of unification and friendly competition, Confederate merchandise becomes normalized as an expression of market choice rather than a representation of the white supremacist, slave-holding South. The banality of such Confederate merchandise was evident to us when a guide told of a visiting boy and his "little Confederate hat." We were sitting in the basement of a building that once housed Confederate sharpshooters and now acts as tourist lodging. As she told the story,

Earlier that day, the boy toured an orphanage. He was an only child and asked that, if there were any ghosts, could they "please come home with me." After leaving Gettysburg, the boy and his mother stopped at a diner. There, his mother took a picture of him and found a ghostly presence. When describing the picture, the guide adds that the boy is wearing his "little Confederate hat." She passes her phone to show us the image. There does appear a ghostly image of a child near him. Yet, what feels more striking—even haunting—is the Confederate flag blazoned atop his head.

With the reconciliation narrative, there is friendly tension, an ongoing war—yet one that is largely commercialized, deracialized, and depoliticized. With the array of shops and items, it feels like a theme park—like Salem, New Orleans, and the Myrtles Plantation—where one can dress up as their favorite hero or villain, depending on what side they choose. This certainly sells products by appealing to the broadest range of (white) customers. Indeed, reconciliation was grounded on white federal officials' and citizens' desires to not upset the "fragile peace" in order to present national healing and "reconciliation among white Americans" (Williams, 2023, 248). And, when, glorifying both sides, we see a sharper form of haunting, that of the lost cause, bolstered by white supremacy not so extinguished with the abolition of slavery.

Lost Cause Narrative

In *Black Reconstruction*, W.E.B. Du Bois (1935, 715–716) hits on a key aspect of the lost cause myth—that the South fought for liberty by asserting state's rights rather than to maintain slavery, which, as the myth has it, had been gradually decreasing in significance:

> Of all historic facts there can be none clearer than that for four long and fearful years the South fought to perpetuate human slavery; and that the nation which "rose so bright and fair and died so pure of stain" was one that had a perfect right to be ashamed of its birth and glad of its death. Yet one monument in North Carolina achieves the impossible by recording of Confederate soldiers: "They died fighting for liberty!"

The lost cause also presents the South as having put on an "admirable struggle against hopeless odds" (Gallagher, 2008, 2) and that General Robert E. Lee was one of the greatest generals in history. Confederate General Jubal Early's former staff officer John W. Daniel wrote in 1880 of Lee as a godlike figure, "The Divinity in [Lee's] bosom shone translucent through the man and his spirit rose up to the Godlike" (Freeman, 1934–35). The *Encyclopedia of American Biography* (Garrity, 1974) describes Lee as "a rare example of a man who looked like a perfect soldier" (see Bonekemper, 2015, 111). Consequently, Confederate General James Longstreet is seen as responsible for the

Confederate loss at Gettysburg and, thus, the Civil War. Meanwhile, Union General Ulysses S. Grant is considered "an incompetent 'butcher' who won the war only by brute force and superior numbers." Gettysburg was, as Kenneth Nivison (2012, 293) describes of the myth, "a Union victory with Confederate valor" (Bonekemper, 2015, 150). While these assertions have been continuously rejected by historians,[6] the lost cause narrative has persisted—and is present in the ghost tourism industry.

The lost cause narrative is perhaps most prominent through the figure of the Confederate ghost. While touring Gettysburg, we were introduced to a cast of Confederate characters. There is Walter who haunts the Farnsworth House Inn; he has a red mustache and a predilection for pulling on women's hair. Similarly, Albert Wood lives in the upstairs of the Ghosts of Gettysburg headquarters. Nameless Confederate soldiers, too, abound: a red-haired and blue-eyed Confederate soldier was said to be seen in the 1970s in the backseat of the car of a young man en route to a high school dance; Confederate soldiers have been known to haunt the hallways of the Adams County Library; and a Confederate ghost appeared to John Burns, the then-mayor who took up arms with the Union. In one instance, a guide initiated an interaction with a Confederate soldier named "Tennessee." We were on a bus tour that stopped at Sach's Bridge, where, as ghost legend has it, three Confederates were hanged for being spies (Schlosser, n.d.). Our guide directed the group to a rail next to an accessibility ramp. We recorded of his demonstration.

> Upon request from the guide, a white, possibly thirty-something, man offers a lit cigarette. After several drags, the volunteer places it on the rail, as instructed. We wait. A child asks the prospective ghost if we can communicate with him, referring to him as "Tallahassee," which prompted laughter.
>
> The embers light up, and the guide confidently points out that it must be Tennessee taking a drag. The cigarette soon falls, the guide confirming that the ghost must have whacked it. The volunteer crushes the slightly used cigarette among a collection of butts on the ground; he looks somewhat disappointed, perhaps that he lost a cigarette to what felt, at least to us, like a rather weak demonstration. We realize, too, that as a group we just shared a smoke with a Confederate soldier.

What is the significance of so many Confederate ghosts in Gettysburg? While books and guides have explained the staggering number as related to improper burial and record keeping,[7] Confederate ghosts also hold symbolic power. Apparitions, Mark Nesbitt (2014, 6) explains, have experienced premature death and just want "to be noticed":

> In fact, of all the stories of ghosts I've collected, none has ever harmed a living human being.
>
> So, it seems that in virtually all the cases, all "they" want is to be noticed. And this is very sad.

While ghosts of Confederate soldiers may represent an individual who succumbed to premature death, and on land not of his own (disrupting a desired "good death" (Faust, 2008)), such Confederate ghost sightings also imply larger national meaning—of a Confederate government that has felt a "premature death" and just wants "to be noticed." When Confederate ghosts are spread throughout the land, as characterized in ghost stories, they afford a sympathetic view, not just of the Confederate soldier, but of the Confederacy's lost cause—as well as the ongoing threat of the Confederacy upon war's end.

Indeed, the Confederate dead were quite threatening in the aftermath of the war, as KKK members dressed up in sheets or other makeshift costumes during nightly raids to Black people's homes, claiming to be the ghosts of the Confederate dead (Dickey, 2016, 209–210). This does not mean those targeted actually believed them to be ghosts. Lorenza Ezell explained of being targeted, as told by Kidada Williams (2023, 250),

> The first time Lorenza's family was struck [in 1868], men dressed in sheets went to his mother's house at midnight, claiming they were soldiers who had died in the war and had returned from the dead. The family was alarmed by the raid, but not because they were afraid of ghosts. "My mama never did take up no truck with spirits," Lorenza said, "so she knew it was just a man."

Nevertheless, the haunting imagery of the Confederate dead coming to life in ghost form symbolizes that the slave regime lives on. As Gladys-Marie Fry (2001, 136) points out,

> The concept of the returning Confederate dead was meant to suggest that the slave regime had not ended, though the South was subdued, and that former controls were still being perpetuated. . . . No longer limited by mortal flesh, the ghosts of the Confederate dead could follow freedman anywhere, appearing at any time, any place.

Thus, when guides recount the ghostly sprawls of Confederate soldiers still haunting the land, they also tell of the *haunting*—and very manifest—reality of white supremacy not wholly extinguished with the abolition of slavery.

In this regard, the Confederate ghosts represent a battle not yet finished. In ghost stories, the Confederacy is never totally defeated, as it lives on in the afterlife. At Gettysburg College, our guide told us of Robert E. Lee still commanding troops in the college's Glatfelter bell tower. Mark Nesbitt and Patti Wilson (2008, 35) furthermore write of the phantom North Carolina brigade marching in the afterlife, at times disguised by the mists of the farm fields:

> Once the summer sun begins to set, the farm fields around Gettysburg have always generated mists that seem to hang, head-high, in long, silver lines,

looking, for all the world, like phantom brigades in their battle lines, advancing once again. Apparently, some of these mists out on the Forney Farm are more animated than others. Are the phantom soldiers still advancing?

Confederate ghosts—as depicting the *spirit* of the Confederacy—thus fight on. A T-shirt sold at a tourist shop displays such messaging. It has three ghostly figures on it screaming, with text that reads, "The War Wages On . . ."

Confederate merchandise sold in tourist shops does much to bolster the lost cause narrative. Confederate memorabilia, or what Tony Horwitz (1998, 31) refers to as "Confederate Kitsch," is found in a number of shops, alongside ghost books and T-shirts. The Confederate flag appears on stickers, patches, coffee mugs, shot glasses, pencils, license plates, can coolers, soda, bandanas, playing cards, and fleece blankets, and it is emblazoned on many T-shirts, made in Haiti, China, India, and the United States. Such items sit next to politically conservative "MAGA" and "Back the Blue" merchandise. Numerous T-shirts echo the lost cause myth, reading:

"If This Flag Offends You, You Need a History Lesson"
"Protect Our Flag"
"One Yank at a Time"
"Heritage, Not Hate"

Confederate merchandise includes ghost products as well. One T-shirt reads, "A Ghost Town with a History Problem," and another declares, "Past Lives Matter," a clear subversion of #BlackLivesMatter (Figure 4.1).

To be sure, pro-Confederate merchandise aligns with broader Confederate memorialization, as the ghost tourism industry is a part of heritage tourism. When visiting Cemetery Ridge, where over 6,000 Confederate soldiers were killed during Pickett's Charge, we spotted a miniature Confederate Flag and Confederate Currency, both purchasable at the local shops (Figure 4.2). They sat next to the death marker of Confederate General Lewis Armistead, who had broken the Union lines yelling, "We must give them the cold steel! Who will follow me?" His is one of 1,328 monuments, memorials, markers, and plaques on the battlefield, with roughly a quarter of them being Confederate (National Park Service, 2022). The Virginia Monument is the most visited and includes Robert E. Lee atop a horse peering over Pickett's Charge with text reading, "Virginia to her sons at Gettysburg." Although such memorials may appear to merely provide historical documentation, they "arrest time in a particular space" (Young, 2022, 240), much like the Confederate ghosts haunting the land. And, as James Loewen (2019) suggests, "[W]hat a community erects on its historical landscape not only sums up its view of the past but also influences its possible futures." With Confederate sympathy still haunting the landscape, the future of racial justice appears grim.

FIGURE 4.1 "Past Lives Matter" T-Shirt

It is important to consider, too, how such depoliticization and deracialization within the ghost tourism industry also omits a more critical view of the North. What is suppressed from historical memory when the North is not critically considered in the ghost tourism industry?

What about the North?

A dominant assumption, which goes unchallenged when slavery is not discussed in the ghost tourism industry, is that the North fought with the primary objective of abolishing slavery. Rather, the North sought, first and foremost, to preserve the Union without, as W.E.B. Du Bois (1935, 716) puts it, "the slightest idea of freeing the slave." In 1864, Fredrick Douglass made this point in his "Mission of the War" speech, proclaiming that the North was "like the south, fighting for National unity; a unity of which the great principles of liberty and equality, and not slavery and class superiority, are the corner stone" (Serwer, 2022, 227). Douglass, furthermore, critiqued Lincoln for his considering a plan of colonization, which would have called for sending millions of emancipated enslaved people to Africa (Hannah-Jones, 2021, 22).

FIGURE 4.2 Memorial at the Armistead Marker

The 1863 Emancipation Proclamation, which declared "that all persons held as slaves" within the Confederacy "are, and henceforth shall be free," was also not always met with enthusiasm in the North, including by Union soldiers. One Massachusetts artilleryman stated, "I am a strong union man . . . but I am not willing to shed one drop of blood to fight Slavery up or down" (Faust, 2008, 25). When Lincoln authorized the recruitment of Black troops by 1862, they were treated poorly in the Union ranks. As James McPherson (1988, 621) describes,

> The black regiments reflected the Jim Crow mores of the society that reluctantly accepted them: they were segregated, given less pay than white soldiers, commanded by white officers, some of whom regarded their men as "n*****s," and intended for use mainly as garrison and labor battalions
>
> *(Serwer, 2022, 227).*[8]

And, after the war, the Thirteenth Amendment barely had enough votes to be ratified in Congress, even without Confederate involvement in the government (Williams, 2023, xvii).

Certainly, the preservation of the Union, with or without slavery, also meant the continuation of white settler colonialism. Gettysburg was settled by James Gettys, a farmer's son who was part of the German and Lutherans and Scots-Irish Presbyterians settlement. Getty bought 116 acres from his father. It was an apt location for trade, as "the principal-north road to Harrisburg crossed the east-west road heading toward South Mountain and the Cashtown Gap" (Guelzo, 2013, 4). This threatened the Seneca people at the river's edge, the Conestoga tribe along the Susquehanna River, and the Lenni-Lenape tribe east of Gettysburg (Silva, 2022).[9] Given that 10 major roads led through Gettysburg, this white settler colonial geography culminated in the clash between Union and Confederate soldiers (Christ, 1997), producing the "battle by accident" (Catton, 1963, 36). In addition, Union General George Meade, who won the Gettysburg campaign, fought in the Second Seminole War (1835–1842) and was a staff officer in the Mexican-American War (1848). Both conflicts were paramount for U.S. colonial expansion across Indigenous land.[10]

The preservation of the Union atop colonized land also meant the development of free market racial capitalism (Robinson, 2020). It should be noted that the industrializing North had indeed benefited from slavery by relying on cotton and other Southern goods for mass production, and cities such as Boston, Philadelphia, and New York were dependent on the slave economy (Manjapra, 2022, 11). Yet, by the time of the Civil War, "Monopoly capitalism was on its way," as C.L.R. James (2012, 61) has put it:

> With increasing industrial expansion of the North, that domination was now in danger. Both North and South were expanding westward. Should the new states be based on slavery as the South wanted or on free capitalism as the North wanted? This was not a moral question
>
> *(57).*

And, once freed, James continues, the North "left the negro to his fate. . . . Landless, his Northern collaborators gone, he was whipped back to an existence bordering on servitude" (63). Indeed, the federal government did little to protect the newly freed from violent white retribution and from state sanctioned convict leasing and sharecropping (Williams, 2023). These forms of social control would take over the North and the South through Jim Crow segregation and mass incarceration. As Viviane Saleh-Hanna (2015, para. 5) contends,

> Though seemingly diametrically opposed, each White side of this bloody tale stands firmly rooted in anti-Blackness driven and legitimated by their own images of White superiority. On one side of White colonial's coin stand slaveholders and their plantations built on stolen lands hanging on, by any means necessary, to a White supremacist slave economy of anti-Black exploitation. On the other side of capitalism's racist coin stand White (self-proclaimed) anti-slavery abolitionists and their criminal justice system built upon a stolen sense of justice hijacked and replaced by imperialist and racialized constructions of crime and criminality.

When deracialized, another aspect of the Civil War is not discussed—that of the aforementioned Black soldiers, despite that 10 percent of Union soldiers were indeed Black. Importantly, there were no Black soldiers in Gettysburg, as those recruited in the fall of 1862 needed training. Nevertheless, a general whitewashing of soldiers excludes from historical memory Black soldiers' fight for emancipation.[11] In May 1861, Douglass argued, "Once black soldiers were given a chance to liberate themselves, they provided a powerful symbol for the deliverance of the American people through warfare" (paraphrased by Nudelman, 2015, 10).[12] Sojourner Truth, too, recruited Black soldiers for the Union, remarking to her friend Samuel Rogers with characteristic wit, "Just as it was when I was a slave—the n*****s always have to clean up after white folks" (Fitch & Mandziuk, 1997, 35). On June 2, 1863, one month before the Gettysburg battle, Harriet Tubman led a major military operation during the Combahee Ferry Raid in South Carolina. She directed 150 Black soldiers to rescue more than 700 enslaved people. This "bankrupt[ed] a network of rice plantations that were funding the Confederate effort" (Gumbs, 2023, 15).

In confronting the legacy of slavery in the South, it is, thus, important to consider how it was instrumental to the rise of the industrial North and how its legacy carries on in both the North and the South. It is this legacy, as well as resistance to it, that we continue to explore in the final chapter.

Conclusion—the Haunting of the Confederacy

We were at an outside beer garden located at a prominent restaurant in town when we wrote the following note. It was a Friday evening, and many were out socializing. It was also the day the U.S. Supreme Court overturned *Roe v. Wade*. We overheard the conversation of a group of middle-aged white people sitting near the outside bar.

> As they drink White Claws, one woman extends approval of the SCOTUS decision, "It's about time," she states.
>
> The conversation turns to race. She continues, "the North should have never won the Civil War. These people are tearing up our streets, murdering and raping our white women." A man stands up and heads to the table, adding, "Blacks are bad people, and honest people need to take back our country."
>
> At the same time, we look to the street and see a guide dressed as Robert E. Lee gathering participants for a ghost tour (Figure 4.3).

While certainly not a monolithic sentiment in the town (and the restaurant apologized for such rhetoric when we called),[13] such an ethos of white supremacy is cultivated when Confederate sympathy is maintained and promoted within the ghost tourism industry—in the town and abroad. Indeed, a professor at University of Stuttgart, Germany, Wolfgang Hochbruck, speculated to Tony Horwitz (1998, 187),

FIGURE 4.3 Tour Guide Dressed as Robert E. Lee

> I think some of the Confederate reenactors in Germany are acting out Nazi
> fantasies of racial superiority. They are obsessed with your war because
> they cannot celebrate their own vanquished racists.

In this regard, the material and ideological conditions of the Civil War
remain intact—the social violence left unresolved (Gordon, 2011). The
atrocity, reconciliation, and lost cause narratives, as told within the ghost
tourism industry, align with a historical denial of the role of slavery in the
Civil War (Cohen, 2001). While a haunted space may be due to the sheer
number of deaths accrued, there is something more to this—something that
persists and sits uncomfortably within the landscape, something "raw and
unresolved" (Horwitz, 1998, 386). "[T]he story of slavery and emancipation," as Kris Manjapra (2022, 9) puts it, "is not a story of endings, but of
unendings." As such, we cannot reconcile with America's horror stories—of
witch hunts, white settler colonialism, chattel slavery, and ongoing
gendered and racialized control—unless we confront their haunting presence
today.

Notes

1 Peter Manseau (2017, 135–137) explains of Alexander Gardner photographs of Confederate soldiers lying along the Hagerstown Pike on September 17, 1862: "Gardner positioned his camera so that the fence stretched diagonally across the frame, seeming to disappear into the distance. The bodies, bloated in the heat, stiff with rigor mortis, seemed ready to rise and walk away. One man's arm stretched into the air as if he'd died while reaching for a rope with which to pull himself out of harm's way. Some photographers might have made the dead the only detail worth noting in the picture. Gardner instead directed the camera's gaze at five sections of posts and split rails. The fence became the perfect focal point for an image ultimately about a nation divided."

2 The combined rate of death is six times that of World War II. The death toll of 630,000 was the equivalent to six million in 2008 (Faust, 2008).

3 Bruce Catton (2013, 84) identifies weaponry as contributing to the death toll. Soldiers relied on close contact fighting and often resorted to bayonet use. Minie balls shattered bones upon contact, and artillery canons shot canisters that "consisted simply of a tin can full of lead slugs somewhat similar than golf balls."

4 Mark Nesbitt and Patty Wilson (2008, 277) describe field hospitals in *The Big Book of Pennsylvania Ghost Stories*, "If there ever was a hell on earth, it was a Civil War hospital. Apprehensive soldiers were brought into the hospital, where they observed other men with hideous wounds and in every degree of pain imaginable. The floors were slick with blood and gore. The wounded sat along the perimeter while the surgeons hunched over makeshift operating tables in the center—perhaps a door taken off its hinges and laid across two sawhorses. The soldiers lay on the tables half-naked, spouting blood from a leg or arm, or perhaps emanating the foulest of smells from a gut wound. But abdominal surgery was almost unknown, and they might be sewn up and left in a corner to die."

5 Tony Horwitz (1998, 187) accounts for banter that occurred between professor Wolfgang Hochbruck, who was dressed in a Union uniform, and a man wearing a Confederate kepi: 'As we left the graveyard, a man in a Confederate kepi spotted Wolfgang and shook a fist, shouting in mock fury, "We'll git you next time, Yank!." "Oh yes?" Wolfgang replied, playing along. "At Gettysburg?"'

6 Edward Bonekemper (2015) dispels several aspects of the Lost Cause Myth. Slavery was booming in 1850 and Southern states wanted stronger federal oversight of Northern states to return escaped enslaved people. The importance of slavery was clear in South Carolina's December 24, 1860 Declaration of Secession which read, "[A]n increasing hostility on the part of the non-slaveholding States to the institution of slavery, has led to a disregard of their obligations, and the laws of the General Government have ceased to effect the objects of the Constitution." While certainly individual Confederate soldiers may have been ambivalent over slavery, over one third owned slaves, or were "the sons of households owning slaves," and more than half of the Confederate officers owned slaves (Guelzo, 2013, 15).

Regarding the valorization of Lee, he did make a series of tactical errors, including self-defeating aggressiveness. He held a one-stage theater in Virginia, with much of the Civil War fought there, and he did not execute orders clearly. Lee also made mistake with the Pickett's Charge command and had even "ordered George Pickett to destroy his scathing report on the disastrous charge that bore his name" (Bonekemper, 2015, 182). Longstreet, who was considered Lee's "workhorse," also opposed Lee's tactics in Gettysburg. Lee admitted as much, as Bruce Catton (1963, 102) confirms, "It should be noted that General Lee—who, after all, was in the best position to judge—never uttered a word of complaint about Longstreet's behavior on this day." Catton (130) continues, "[Lee] had refused to put up any alibis for the defeat, taking all of the blame upon his own shoulders."

Furthermore, while Union soldiers did outnumber Confederate soldiers (1 to 3.5), Grant did show military prowess. He had topographical memory and communicated well. Grant was able to outmaneuver and capture the fortified city of Vicksburg in the summer of 1863, which elevated him to general-in-chief of the Union Army. The Union victories in Gettysburg and Vicksburg meant, Catton (1963, 131) concludes, "that it was no longer possible for the Confederacy to win the war."

7 A guide explained that the rate of Confederate ghosts was due to Gettysburg residents destroying Confederate records as a response to the federal government not paying for their burial. Nesbitt and Wilson (2008, 234) argue that Confederate ghosts abound from improper burial, stating, "Unconsecrated burials are said to produce hauntings. The North Carolinians, with nothing more than a few inches of Pennsylvania loam scraped over them, lay in uneasy sepulchers."

8 The legacy of white retribution toward Black soldiers was prevalent almost a century later when WW II Black army sergeant Isaac Woodard was pulled from a Greyhound bus and attacked by a South Carolina police officer in 1946. The officer beat him so badly that it ruptured his eyes which left him blind. Woodard was in uniform at the time (Rickford, 2022, 314).

9 Gettysburg College has done much to resurface this history through a Land Acknowledgment Committee and student-run group, Students for Indigenous Awareness (SIA) (Silva, 2022).

10 Speaking to Western expansion, Lincoln approved in 1862 the hanging of 38 Dakota men in Mankato, Minnesota, as a result of the Dakota War. As Scott Berg (2013, 224) describes of the hangings, "Eager spectators were staking claims on rooftops and balconies and windows, in the streets and up on the bluff overlooking town, on the sandbar in the river and even along the opposite shore."

11 Of note, several thousand enslaved people performed combatant duties in the Army of Northern Virginia and Black laborers also accompanied the Army of the Potomac (Gallagher, 2008, 58). They are also erased in the tourist industry as well as in popular media.

12 It should be noted that Black people fighting the war as soldiers was not unanimously supported by Black Americans. W.E.B. Du Bois (1935, 110) points out the cultural valorization of soldiers: "How extraordinary, and what a tribute to ignorance and religious hypocrisy, is the fact that in the minds of most people, even those of liberals, only murder makes men. The slave pleaded; he was humble; he protected the women of the South, and the world ignored him. The slave killed white men; and behold, he was a man!" (Faust, 2008, 48).

In The Will to Change, bell hooks (2004, 170) further criticizes the "war mentality" espoused through masculine socialization. "The warrior's way," she reminds, "wounds boys and men; it has been the arrow shot through the heart of their humanity. . . . Even though war is failing as a strategy for sustaining life and creating safety, our nation's leaders force us into battle, giving new life to the dying patriarchy."

13 Of note, several businesses did display LGBTQIA+ Pride Flags, perhaps given that we visited on the Pride Month of June. All of this is to say that Gettysburg is not monolithic, and there appears in the town a divide exemplified in the larger nation between more conservative and more progressive ideals.

References

Adams, Henry. 1880. *Exoduster Testimony*, Part II, 113.

Berg, Scott W. 2013. *Nooses: Lincoln, Little Crow, and the Beginning of the Frontier's End*. New York: Vintage.

Bonekemper, Edward H. 2015. *The Myth of the Lost Cause: Why the South Fought the Civil War and Why the North Won*. New York: Simon & Schuster.

Brazile, Donna. "The U.S.Constitution." In *Four Hundred Souls: A Community History of African America, 1619–2019*, edited by Ibram X. Kendi and Keisha N. Blain, 153–157. New York: One World.

Catton, Bruce. 1963. *The Battle of Gettysburg*. New Word City. New York: American Heritage Publishing Co. Inc.

Catton, Bruce. 2013. *Gettysburg: The Final Fury*. New York: Vintage.

Christ, Elwood W. 1997. "Building a Battle Site: Roads to and through Gettysburg." *Adams County History*, 3(1): 41–70.

Cohen, Stanley. 2001. *States of Denial: Knowing about Atrocities and Suffering*. Hoboken: John Wiley & Sons.

Conservation Fund. "Gettysburg National Military Park." https://www.conservationfund.org.

Dickey, Colin. 2016. *Ghostland: An American History in Haunted Places*. Westminster: Penguin.

Diouf, Sylviane A. 2022. "Maroons and marronage." In *Four Hundred Souls: A Community History of African America, 1619–2019*, edited by Ibram X. Kendi and Keisha N. Blain, 89–92. New York: One World.

Du Bois, W.E.B. 1935. *Black Reconstruction in America: An Essay Toward a History of the Part Which Black Folk Played in the Attempt to Reconstruct Democracy in America, 1860–1880*. A. Saifer

Faust, Drew Gilpin. 2008. *This Republic of Suffering: Death and the American Civil War*. New York: Vintage.

Freeman, Douglas Southall. 1934–1935. *R.E. Lee* (vol. IV). New York: Charles Scribner's Sons.

Fitch, Suzanne P., and Roseann Mandziuk. 1997. *Sojourner Truth as Orator: Wit, Story, and Song*. New York: Bloomsbury Publishing.

Fry, Gladys-Marie. 2001. *Night Riders in Black Folk History*. Chapel Hill: University of North Carolina Press Books.

Gallagher, Gary W. 2008. *Causes Won, Lost, and Forgotten: How Hollywood and Popular Art Shape What We Know about the Civil War*. Chapel Hill: University of North Carolina Press.

Garritty, John A. 1974. *Encyclopedia of American Biography*. New York: Harper & Row.

Gordon, Avery. 2011. "Some Thoughts on Haunting and Futurity." *Borderlands*, 10(2): 1–21.

Guelzo, Allen C. 2013. *Gettysburg: The Last Invasion*. New York: Vintage.

Gumbs, Alexis Pauline. 2023. "Learning to Listen." In *Healing Justice Lineages: Dreaming at the Crossroads of Liberation, Collective Care, and Safety*, edited by Cara Page and Erica Woodland, 15–20. Berkeley: North Atlantic Books.

Hannah-Jones, Nikole. 2021. *The 1619 Project: A New Origin Story*. New York: One World.

hooks, bell. 2004. *The Will to Change: Men, Masculinity, and Love*. New York: Atria Books.

Horwitz, Tony. 1998. *Confederates in the Attic: Dispatches from the Unfinished Civil War*. New York: Vintage.

James, C.L.R. 2012. *History of Pan-African Revolt*. Binghamton: PM Press.

Jones, Martha S. 2022. "The American Revolution." In *Four Hundred Souls: A Community History of African America, 1619–2019*, edited by Ibram X. Kendi and Keisha N. Blain, 139–142. New York: One World.

Loewen, James W. 2019. *Lies Across America: What Our Historic Sites Get Wrong*. New York: The New Press.

Manjapra, Kris. 2022. *Black Ghost of Empire: The Long Death of Slavery and the Failure of Emancipation*. New York: Simon & Schuster.

Manseau, Peter. 2017. *The Apparitionists: A Tale of Phantoms, Fraud, Photography, and the Man Who Captured Lincoln's Ghost*. Boston: Houghton Mifflin Harcourt.

Mark Nesbitt's Ghosts of Gettysburg, 2022. "About Ghosts of Gettysburg." https:// ghostsofgettysburg.com.

McPherson, James M. 1988. *Battle Cry of Freedom.* New York: Oxford University Press.

National Park Service. 2022. "*Confederate Monuments.*" https://www.nps.gov.

Nesbitt, Mark. 2012. *Ghosts of Gettysburg: Spirits, Apparitions, and Haunted Places of the Battlefield.* Gettysburg: Second Chance Publications.

Nesbitt, Mark. 2014. *Ghosts of Gettysburg III: Spirits, Apparitions, and Haunted Places of the Battlefield.* Gettysburg: Second Chance Publications.

Nesbitt, Mark, and Patty A.Wilson. 2008. *The Big Book of Pennsylvania Ghost Stories.* Mechanicsburg: Stackpole Books.

Nivison, Kenneth. 2012. "Gettysburg and the Americanization of the Civil War." In *The Battlefield and Beyond: Essays on the American Civil War,* edited by Clayton E. Jewett. Baton Rouge: Louisiana State University Press.

Nudelman, Franny. 2015. *John Brown's Body: Slavery, Violence, and the Culture of War.* Chapel Hill: University of North Carolina Press Books.

Phelps, Stuart Elizabeth. 1868. *The Gates Ajar.* Boston: Fields, Osgood & Co.

Rickford, Russell. "The Black Left." In *Four Hundred Souls: A Community History of African America, 1619–2019,* edited by Ibram X. Kendi and Keisha N. Blain, 312–316. New York: One World.

Riley, Johlene. 2011. *Ghostly Encounters of Gettysburg.* Gettysburg: Arbor House.

Riley, Johlene. 2014. *Ghost Hunting—The Gettysburg Files.* Gettysburg: Arbor House.

Robinson, Cedric J. 2020. *Black Marxism, Revised and Updated Third Edition: The Making of the Black Radical Tradition* (ed.) Chapel Hill: University of North Carolina Press Books.

Saleh-Hanna, Viviane. 2015. "Black Feminist Hauntology. Rememory the Ghosts of Abolition?"*Champ Pénal/Penal Field.* 12.

Schlosser, S.E. "Haunted Sach's Bridge." *S.E. Schlosser. https://www.seschlosser.com.*

Serwer, Adam. 2022. "Frederick Douglass." In *Four Hundred Souls: A Community History of African America, 1619–2019,* edited by Ibram X. Kendi and Keisha N. Blain, 225–229. New York: One World.

Silva, Katelyn. 2022, June 1. "Honoring the Indigenous Past and Present: The Land Acknowledgment Statement Formally Recognizes the Land on Which Gettysburg College Sits Is Indigenous Land." *Gettysburg College.* https://www.gettysburg.edu.

Williams, Kidada E. 2023. *I Saw Death Coming: A History of Terror and Survival in the War against Reconstruction.* New York: Bloomsbury Publishing.

Young, Alison. 2022. "The Time of Ghosts: Sites of Violence, Environments of Memory." In *Ghost Criminology: The Afterlife of Crime and Punishment,* edited by Michael Fiddler, Theo Kindynis, and Travis Linnemann, 227–252. New York: New York University Press.

CONCLUSION

Toward a Critical Ghost Tourism

Introduction

We return to the Greater Philadelphia area from our ghost tour trip to Salem, Massachusetts; New Orleans and St. Francisville, Louisiana; and Gettysburg, Pennsylvania. From our trip, we identified numerous themes, primarily that, in appealing to white consumers, ghost tour guides often downplay or omit the realities of colonialism and slavery. In Gettysburg, guides did not discuss slavery as the impetus of Southern secession, and in New Orleans, the story of Madame Delphine LaLaurie's brutality toward those she enslaved implied that slavery was overall humane in the city.

We also saw in stories a constellation of racist stereotypes. Enslaved Chloe at the Myrtles Plantation is considered a "mistress" and "conniving," and Tituba, or Tatebe, is presented in Salem as practicing Voodoo and speaking about the devil to the white Puritan children. Punishment, too, is marketed through stories of tortured enslaved people, Black women suffering, and ghost stories of white Puritans, such as Bridget Bishop, who was targeted for witchcraft, and who is said to haunt local restaurants and hotels to entice customer patronage. We also saw the appropriation of Voodoo, Vodou, or Vodun, such cooptation stripping the revolutionary spirit of a belief system born out of slavery for whitewashed self-betterment. Such whitewashing allows audiences to delve into the dark history of the United States, without acknowledging the horrors of colonialism or slavery, which still haunt the land.

Yet, we also tracked these American horrors by looking at markers of slavery, of the French fleur-de-lis symbol in New Orleans that was branded on enslaved people, or the everyday donning of the symbol of the Confederate flag in Gettysburg, which conceals the brutality of the Southern slave system. Chloe's narrative also depicts—and makes visible more generally—sexual

DOI: 10.4324/9781003397809-6

violence toward enslaved Black women, which extends to the contemporary criminal punishment system—as Black women are deemed as having no "self" to defend (Kaba, 2019). One New Orleans guide, a young Black woman, linked abuses the LaLauries perpetrated on those they enslaved with contemporary racism in the medical establishment. She offered of her experience, "When I go to the hospital, I am just prescribed Tylenol" (see Washington, 2008).

We noted resistance too. Within the Massachusetts Bay Colony, Indigenous people fought white settler colonialism through violent force, and Tatebe utilized a variety of techniques to protect herself when falsely confessing to practicing witchcraft, thus affording a "new idiom of resistance" (Breslaw, 1995, xx). Chloe, too, maintained secrets for her survival, and Voudou was essential for the 1791–1804 Haitian revolution, which inspired further revolts such as the 1811 German Uprising. Lesser known figure, Voodoo priestess Besty Toledano, as we also documented, advocated for those in court who were persecuted for engaging in Voodoo ceremonies (Long, 2007, 210). And, during the Civil War, oft-ignored Black soldiers took up arms to fight for the Union, while Harriet Tubman led a major military operation during the Combahee River Raid, which rescued 700 enslaved people (Gumbs, 2023). "Time and again white racism produced Black resistance," Herb Boyd (2022, 84) puts it. "It is one of the longest-running plotlines in African American history."

While such atrocity and resistance is obscured or redacted in the ghost tourism industry, we have argued that the ghost(ed) offers a way to see unresolved social violence that haunts the past-present-future (Fiddler et al., 2024). Ghosts, thus, are not going away any time soon, whether as a product to allure tourists or as a powerful metaphor for unresolved social violence (Gordon, 2011). Nor should they. As with Toni Morrison's (1987) *Beloved*, the ghost holds a place to remember and reckon with past atrocities. Tiya Miles (2015, 126) puts it well:

> In African American tales of haunting, spirts of the dead are neither frivolous nor romantic. Their presence signals a need for those in the present to deal directly with a dangerous past that will not rest.

Indeed, wrestling with America's haunting past beckons us to "remake good relations," as Kris Manjapra (2022, 10) further contends,

> This haunting beckons us to remake good relations from within the ongoing aftermath of historical trauma. The ghosts in our history demand reparative action—diverse practices of reparations, restitution, and redress—by all of us standing on the ground of slavery and by the ruling order built upon it.

In this regard, *ghosts do good work*. In our concluding remarks, we consider how scholars and activists have conjured ghosts in history, memorialization, and movement. Recognizing these areas of ghost(ed) social justice is key to a critical ghost tourism, as we outline in the final section.

Conjuring Ghosts in History

To seek ghosts of past atrocity, it is imperative to engage with narratives of those who have endured white supremacy's violence. We again cite the words of W.E.B. Du Bois (1935, 715) who recommends accessing slave biographies to counter dominant narratives, such as the antebellum splendor narrative we cited at the Myrtles Plantation:

> Shall we accept the conventional story of the old slave plantation and its owner's fine, aristocratic life of cultured leisure? Or shall we note slave biographies, like those of Charles Ball, Sojourner Truth, Harriet Tubman, and Frederick Douglass; the careful observations of Olmsted and the indictment of Hinton Helper?

Afro-narrative storyteller Elizabeth James (2014), who has family roots in New Orleans, digs into historical, familial, and personal memory to tell of Madame LaLaurie from her grandmother's perspective. Her grandmother instructed her, when walking past the mansion, to "cross to the other side of the street and make a sign of the cross, both for protection and in honor of the ancestors" (Miles, 2015, 128). Not a faceless, animalistic mass (Foley, 2021), but those who were abused in LaLaurie's residence are garnered here with respect. To gain a more complex and historically accurate understanding of Marie Laveau, Carolyn Long (2007, xvii) also utilizes news reporting, government records, and interviews of those who knew her, as cited in the 1930s and early 1940s Louisiana Writers' Project. In doing so, she gives a more complex image of Laveau, while also recognizing lesser known historical figures, such as voodoo priestess Besty Toledano.

In *I Saw Death Coming*, Kidada Williams (2023) examines testimony of those who faced the terror of KKK night raids during Reconstruction.[1] In 1871, Edward Crosby reported he "felt the thunder of a team of horses" and "saw about thirty disguised men descending on his home, their mounts draped by full-body cloth coverings" (xi). While Crosby certainly feared the violation of his right to vote, it was "the possibility they would return," which, as Williams paraphrases, "presented a far greater danger to his family's well-being and freedom" (xix). As targets of white vigilantes, the newly freed reported being robbed of possessions, disenfranchised, sexually assaulted, and were made to "lie out" or relocate altogether, severing bonds between families and community, thus stifling possibilities for intergenerational wealth. The night raids enacted, she contends, the war after the Civil War, with the South initiating "the pursuit of the Confederate cause by other means" (xvii).[2] And, as Williams continues of the interviewees, those attacked created a codex of survival and "soft blue print of hope" (Miles, 2022) through mutual aid; they stood as lookouts, took care of each other's property, and risked their lives when speaking to government officials about their individual and collective experiences. (128).

Conjuring this history is a collective endeavor. In *Four Hundred Souls* (Kendi & Blain, 2022), 80 Black writers and 10 Black poets commemorated in 2019 the 400th anniversary of August 1619, when the first documented enslaved Africans were brought into Jamestown, Virginia, in the slave ship *White Lion* and who were listed as "Twenty and odd Negroes" (Kendi, 2022, xiii).[3] Each contributor takes on five years at a time. They create a diverse, collective voice, or a choir as Ibram X. Kendi (2022, xv) describes them. For chattel slavery, authors document the arrivals of the *White Lion* in Virginia in 1619 (Hannah-Jones, 2022) and the slave ship *Ruby* in Louisiana on July 16, 1720 (Diouf, 2022); of the slave trade in New York (Turner, 2022); and of the wealth accumulated from the transatlantic slave trade, including by the Royal African Company (Love, 2022); and slave code regulations regarding interracial relationships (Oluo, 2022), Black women's labor (Stevenson, 2022), weapon possession (Jackson, 2022), Northern escape (Owens, 2022), citizenship (Powell, 2022), slave status related to birth (Morgan, 2022), and baptism (Tisby, 2022). Jericho Brown (2022, 35) encapsulates this history, the "history of the wound," in his poem "Upon Arrival." A portion reads,

> Here is the history of the wound: Somebody brought them. Somebody bought them. Though I know who caught them, sold them, bought them, I'd rather focus on their faces, their names.

The writers also portray resistance and resilience. They discuss the complexity of 1676 Bacon's Rebellion (McGhee, 2022); the 1739 Stono Rebellion in South Carolina (Lowery, 2022); the 1811 German Coast Uprising (Smith, 2022b); and of maroon communities sparking fear in planters' minds (Diouf, 2022). They look to African religions, both Abrahamic and indigenous (see Barber II, 2022). Corey D.B. Walker (2022, 93) illustrates of the early 18th century,

> Freedom would have been a sonic cacophony of beats, rhythms, and melodies, clapping and stomping in syncopated time that moved between and beyond purely notational patterns. It would have resembled, reflected, and refracted the stirrings of an Atlantic world in motion.

They record poet Phillis Wheatley writing of sea voyages, marking a separation between herself from her lineage and community (Gumbs, 2022, 133); conscious hip-hop driving Black political engagement (Kitwana, 2022); and the contemporary influence of and backlash toward Black Lives Matter (Garza, 2022). Imani Perry (2022) showcases Black artists who speak to this ongoing resistance, including James Baldwin, Elizabeth Catlett, Chinua Achebe, and Lorraine Hansberry,

> Black life was not mere endurance but a victory of spirit in the form of human complexity, imagination, resistance, breadth, and depth, precisely the resources that were essential for the coming revolutions.

In this regard, these scholars draw, through historical documentation, the "unbroken links" (Stevenson, 2021, 282), or *chains*, between chattel slavery, extra judicial terror, segregation, mass incarceration, mass policing, mass supervision, and concentrated poverty and organized abandonment (Davis, 2003; Gilmore, 2007). Indeed, those subjected to the horrors of convict leasing, who were treated as even more disposable than enslaved people (Blackmon, 2009), were buried in secret graves after being worked to death (Love, 2022, 50)—and which also have souls which need to be seen, felt, and heard (Deyab, 2016).

Conjuring historical ghosts when documenting oppression, resistance, and resilience can involve fiction writing, as Toni Morrison (1987) accomplishes in *Beloved*. Her magical realism (Deyab, 2016, 16) blends fiction and history, as she was inspired by the story of Margaret Garner as well as the horrors of the Middle Passage (Singh, 2021). As Mohammad Deyab (2016, 16) observes, "*Beloved*'s ghost always stays on the borderline between two worlds: reality and fantasy, history and fiction, past and present." Indeed, given that history is "often written from the perspective of the dominant white culture," Morrison contends that "a fictional account of the interior life of a former slave might be more historically 'real' than actual documents" (paraphrased by Deyab, 2016, 15; see Lucie, 2000).

To this, critical race theorists have utilized fiction as a method to explore enduring systemic racism. Through a series of chronicles, Derrick Bell (1987, 7) writes of extant racism with "stories that are not true to explore situations that are real enough but, in their many and contradictory dimensions, defy understanding" (see also Aja Martinez (2020) on counterstory). Black writers have for instance utilized the horror genre to detail the terrors of white supremacy. Horror, as Jordan Peele (2023, viii) puts it, affords "catharsis through entertainment." In Peele's anthology *Out There Screaming*, 19 Black writers utilize horror, science fiction, and fantasy genres to engage with topics such as lynching (Broaddus, 2023), the Freedom Rides (Due, 2023), and modern-day slave labor (Taylor, 2023). Set in 1920s Georgia, P. Djèlí Clark (2020) writes in dark fantasy novel *Ring Shout* of a trio of Black women monster hunters who fight demons summoned by the KKK called "Ku Kluxes." In the science fiction novel *An Unkindness of Ghosts*, Rivers Soloman (2017) takes on racial stratification and rebellion on the spaceship *HSS Matilda*, as it supposedly travels to the "Promised Land."[4]

Of course, fiction and nonfiction writing are best taken together to create a dynamic image of historical figures and events. As we explored in chapter 1, Maryse Condé (1992) offers a harrowing tale of Black struggle through her fictitious characterization of Tituba. She writes of Tituba's mother being hanged after defending herself from sexual advances from her white enslaver—echoing stories of Black women who are treated as though they do not have a "self" worth defending (Kaba, 2019). Such material need not be antagonistic to historical research. In *Tituba: Reluctant Witch of Salem*, Elaine Breslaw (1995) identifies complexity, agency, and resistance in Tituba's life and confession, as based on rich historical documentation (see also Gibson, 2024). Both review historical truths.

Scholars and activists have also conjured the historical ghosts of those condemned as witches for a broader analysis of patriarchal violence and as an anchor for feminist liberation. While those accused in Salem would not have subscribed to any form of witchcraft, the figure of the witch, as Marion Gibson (2024, 151) describes, is "a metaphor for freedom." As Gibson (2024, 243–244) reminds us,

> [M]edieval demonologists and those who followed in their footsteps imagined witches as embodying their worst nightmares, which included expressions of female power, unfettered sexuality, and subversive political intentions.

In Defense of Witches, Mona Chollet (2022, 254), too, speaks of the power and joy evoked through the witch:

> [T]here can be great joy—the joy of audacity, of insolence, of a vital affirmation, of defying faceless authority—in allowing our ideas and imaginations to follow the paths down which these witches' whisperings entice us. Joy in bringing into focus an image of this world that would ensure humanity's well-being through an even-handed pact with nature, not by a Pyrrhic victory over it—this world, where the untrammeled enjoyment of our bodies and our minds would never again be associated with a hellish sabbath.[5]

African grassroots women's movements have also initiated direct-action *against* the witch label, or at least as it is applied by witch hunters. As a form of resistance, African women "'sit' on witch hunters, disrobe in front of them, and perform shaming acts of staged 'incivility'" (Federici, 2018, 81).

Thus, through historical accounts, fiction, and the reclamation of demonized imagery, we can conjure the ghosts of history—the "misrecognized, the disavowed, and the ignored historical presences among us" (Manjapra, 2022, 4). Perhaps ghost "hunting" is not the ideal term after all—with its violent connotations of capture. Rather, when looking to the ghosts of history, we engage in ghost/spirit conjuring to recognize and resist capitalism's violence and its subsequent erasure. Such ghosts, too, are conjured in memorialization.

Conjuring Ghosts in Memorialization

While places of atrocity have been commercialized and exploited, as we saw at the Myrtles Plantation, such sites can also be modeled in ways that critically engage with these atrocities—conjuring the ghosts of the past through memorialization and critical education.

Sitting 30 miles west of New Orleans in the town of Wallace is the Louisiana Whitney Plantation, which is now a museum. Unlike the Myrtles, it focuses exclusively on the lives of enslaved people—and it is the first plantation to do so. The museum includes an exhibit on the North American slave trade, offers tours of slave cabins, and has "a series of angled granite walls engraved with the names of the 107,000 slaves who spent their lives in Louisiana before 1820" (Amsden, 2015). Such an approach is promoted in Kelly McWilliams's (2023, 4) young adult fiction novel, *Your Plantation Prom Is Not OK*. The story's protagonist Harriet Douglas and her historian father run a Louisiana-plantation-turned-enslaved-people's museum. Yet, a white family purchases the property next door with plans to run it as an antebellum-themed wedding venue, much like the Myrtles. Such writing initiates an important discussion on how to properly engage with plantations for historical engagement.

Prison sites can also critically capture the horrors of mass incarceration, as an extension of chattel slavery, lynching, and racial segregation (Davis, 2003). Philadelphia's Eastern State Penitentiary is one such example. The museum does have a "Haunted Halloween" tour that offers a rather sensationalized engagement with the space. Yet, the museum also includes letters and photographs from those imprisoned and holds several exhibits on mass incarceration, including "Prisons Today," which asks the reader, "Have you ever broken the law?" The answer for many is, or probably should be, the exhibit reminds is, "Yes." The exhibit thus troubles the distinction between the "criminal" and so-called law abiding citizen. This is complemented with "The Big Graph," a 16-foot tall, 3,500-plate steel sculpture in the prison's yard. The graph illustrates the growth of U.S. incarceration rates, the racial breakdown of the U.S. prison population, and capital punishment policies worldwide (Eastern State Penitentiary).

Advocates and organizations have also worked to erect monuments that memorialize past atrocity, such as Alabama's Equal Justice Initiative (EJI). In 2015, EJI created the Community Remembrance Project to bring attention to terror lynchings in the South, in which 4,000 were killed by vigilantes between 1877 and 1950 (Stevenson, 2017, 13). Over 80 historical markers have been installed, and the organization has collected soil from 700 lynching sites. In promoting critical historical education, EJI has also received over 900 racial justice essay entries from high school students. The EJI website, too, includes an interactive map of racial terror lynchings in the South. As a member of the Community Remembrance Project Coalition in Chattanooga, Tennessee, puts it of the initiative, "There can be no reconciliation and healing without remembering the past."

In New Orleans, Marcus Brown, who is an artist, activist, and educator at New Orleans Center for Creative Arts (NOCCA), has initiated the "Slavery Trails" installation. Through interactive augmented reality (AR), viewers are able to display "pink virtual ghosts" of enslaved people on their smartphones or smart devices. As such, Brown challenges a general whitewashing of slavery in the city. As he told *PBS News* (Chavez, 2022),

> But the reality is, a lot of these structures were built with slave labor, and a lot of our streets that we walk on in the French Quarter and elsewhere were built in slave labor. We've never addressed it properly.

He declares, "America needs the ghosts of slavery to remind us of how our nation was really built."

To be sure, one need not create a museum or establish markers to memorialize and publicize racial horror. Mamie Till famously held an open casket funeral for her 15-year-old son, Emmett Till, who was murdered by two white men in Mississippi in 1954. His brutalized figure revealed the Southern horror of lynching (Wells, 1892). As Till put it regarding her decision, "Let the people see what they did to my little boy." More recently, R.I.P. (rest in peace) shirts have memorialized those lost to racialized premature death (Gilmore, 2007), while simultaneously evoking images of joy and pain. They are, as Robin Brooks (2018, 180) claims, a constitutive part of the Movement for Black Lives (M4BL). Vigils furthermore allow for solidarity building and offer space for radical ideas to flourish (Hayes & Kaba, 2023, 180). Ultimately, as Karla Holloway (2003, 3–6) recognizes, "Black folk—whose indomitable and full presence articulates the best of this country's spirit, intelligence, and politics—bridge this cultural haunting [of racialized death] with hope, grace, and resilience."

Conjuring Ghosts in Movement

Scholars and activists have also conjured ghosts to promote community-based social justice. In *Healing Justice Lineages* (Page & Woodland. 2023a), Black feminist scholar Erica Woodland (2023b, 81) defines healing justice as building an ecology of care by drawing on spiritual technologies that "our ancestors" adopted "as part of their overall strategies for liberation." These include "knowledge rooted in memory, intuition, and embodiment, which acts as a conduit to transcend time and space." Healing justice activists do not totally reject Western medicine.[6] Rather, they seek to integrate it within community that honors culture, language, and seeks to address generational trauma from slavery and colonialism—through "care, healing, and liberation" (Page & Woodland, 2023c, 265).[7]

And memory is key to this work. As Cara Page and Erica Woodland (2023b, 4) contend, "Healing justice is memory work. The work of reclamation." For instance, Alexis Pauline Gumbs (2023) evokes the spirit of Harriet Tubman navigating fire, sky, water, and earth to guide herself and others to freedom. In honoring Tubman's activism, Gumbs manages the Combahee Throughline Portal, which connects the Combahee Uprising to Black feminism, and she cofacilitated the Combahee Pilgrimage on the 150th anniversary of the raid (Page & Woodland, 2023a, 291).

The Black feminist Combahee River Collective also carries the legacy of Harriet Tubman. The organization operated in Boston from 1974 to 1980 as a response to racial violence perpetrated by white Bostonians. They collectively engaged in actions that included "study, political analysis, protests, campaigns, cultural production, and coalition work around a range of issues, all with the objective of defining and building Black feminism" (Smith, 2022a, 341). They offered direct support in campaigns to free Joan Little and Ella Ellison; initiated political retreats; drew awareness to murdered Black women in Boston, sparked by the 1979 killings of 15-year-old Christine (Chris) Ricketts and 17-year-old Gwendolyn Yvette Stinson; and they laid the groundwork for intersectionality when asserting that "the major systems of oppression are interlocking" (Combahee River Collective Statement, 1977; Taylor, 2017).[8] Their work certainly holds power today with the #SayHerName campaign (Crenshaw, 2024).

Established on April 6, 2014, Harriet's Apothecary has also created in Tubman's name local and national healing spaces modeled after the Mississippi Freedom schools. They do so to "bring resources and practices of wellness, safety, and care outside of the medical industrial complex." As the organization puts it,

> Harriet's Apothecary envisions a world where Black, Indigenous, and People of color have the power, healing, and safety needed to live the lives we desire for ourselves and our communities
>
> *(Harriet's Apothecary, 2024).*

In these efforts, Harriett's Apothecary embraces, as Adaku Utah and Cara Page (2023, 181) proclaim, the "sacred work of loving and building power with Black people." Alexis Pauline Gumbs, the Combahee River Collective, and Harriet's Apothecary, honor Tubman's legacy, taking seriously Toni Morrison's (Lucie, 2000) observation, "If we don't keep in touch with the ancestors . . . we are, in fact, lost" (Deyab, 2016, 14).

Indigenous people have also adopted a variety of spiritual activities to resist white settler colonialism. The Ghost Dance is a practice that began in 1889 and was based on visions of Paiute elder Wovoka "who prophesied the end of European expansion and the reunion of Indigenous peoples with their ancestors" (Woodland, 2023b, 81). It was, as Woodland elaborates, "a spiritual technology to directly intervene on settler expansion and preserve Indigenous cultural traditions" (81). Nearly a century later, the American Indian Movement (AIM) initiated the Trail of Broken Treaties (1972) to memorialize the Trail of Tears. They led caravans from San Francisco, Los Angeles, and Seattle to march on Washington, DC. As President Richard Nixon was out of town upon their arrival, they commenced a weeklong occupation of the Bureau of Indian Affairs (BIA) to bring attention to their cocreated 20-point plan, which integrated political rights and spiritual practices. The 20-point plan

> called for the restoration of treaty making (ended in 1871); the return of millions of acres of land stolen from Indigenous people; abolition of the BIA; and protection of Indigenous spiritual, religious, and cultural practices, among other demands, to reestablish sovereign relationships between tribes and the US government (American Indian Movement, 1972; paraphrased by Woodland, 2023a, 34)

While the plan was rejected, this action did shape Nixon administration policies, including the Indian Self-Determination and Education Assistance Act of 1975 and the reestablishment of tribes' ability to enter contracts directly with the federal government (Woodland, 2023a, 34–35).

What about those with ancestors who fought for such atrocity, such as the descendants of Confederates? Certainly, during the Civil War the Confederacy experienced major loss of life. At the end of Vicksburg and Gettysburg campaigns in 1863, North Carolinian Catherine Edmondston "called at the houses of eight neighbors and found each one in mourning for a lost husband, brother, or son" (Faust, 2008, 188). Virginian Mary Lee felt at the end of the battle "like a ship without a pilot or compass," with no "God at the helm" (192–193). Subsequently, the cult of the lost cause and celebration of Confederate memory, as Drew Faust (193) recognizes, "were in no small part an effort to affirm that the hundreds of thousands of young southern lives had not, in fact, been given in vain." And, for those interviewed by Toni Horwitz (1998, 174), a connection to the Confederacy offered a sense of belonging. As interviewee Elijah put it, "Ultimately, I guess I'm trying to figure out what my place in the big picture is. . . . I am who I am, geographically and politically, because of what happened here."

Yet, it is imperative to recognize the role of slavery as foundational to the Confederacy: to denounce slavery as constitutive of the Confederacy is certainly not engaging with the past but reinforcing the horrors of "what happened here." Horwitz (1998, 44) asked Black preacher Michael King "if there was any way for white Southerners to honor their forebears without insulting his."

> He pondered this for a moment. "Remember your ancestors," he said, "but remember what they fought for too, and recognize it was wrong. Then maybe you can invite me to your Lee and Jackson birthday party. That's the deal."

In *Robert E. Lee and Me*, West Point history professor Ty Seidule (2021, 11) presents his own Southern upbringing and valorization of Robert E. Lee. As he reflects, "To a boy growing up in Virginia, Lee was more than the greatest Virginian; Lee was the greatest human who ever lived." Yet, he would come to reckon with the larger myth of the lost cause, that he had "failed to look at the two issues that today sear [his] soul: treason and slavery" (215). Now, he challenges the valorization of such figures—that maybe we do not need a Lee and Jackson birthday celebration after all.[9]

This also poses a question on nostalgia. While nostalgia can open up critiques of present economic and social conditions (O'Donovan, 2019, 272), it also tends to create "distorted perspectives that misrepresent the past" (Lowenthal, 2015). White nostalgia in particular acts as "an attempt to escape issues of race by downplaying their implications and rejecting the legacy of slavery" (Adamkiewicz, 2016, 13). A distorted sense of the past disconnects historical atrocity from the present, rendering the witch hunts, colonialism, or slavery as simply "how things were back then" (Wright, 2022, 79). As Kai Wright (2022, 77) puts it,

> We see horrible people as exceptional, and their many accomplices as mere captives of their times. We tell ourselves we would contain such wickedness if it arose today, because now we know better. We've learned. In our illusory past, progress has come in decisive and irrevocable strokes.

Yet, as we have reviewed here, the tendrils of the past, in racist ideology and control, continue on, with racist ideas morphing between religious, scientific, and cultural explanations (Roberts, 2022). Indeed, disconnecting the past-present is a form of historical denial (Cohen, 2013), which permits tourists to freely don the antebellum lifestyle or consume punitive spectacles of Black violence. We must *unwill* willed forgetfulness (Cariou, 2006) and seek what David Waldstreicher (2010) refers to as meaningful silences[10] by conjuring ghosts in historical documentation, in memorialization, and in movement. This is key for a critical ghost tourism.

Toward a Critical Ghost Tourism

We conclude *America's Horror Stories* by noting key elements of critical ghost tourism—of social justice, mapping conditions, reflexivity, and a consideration of what it means to travel and return home.

Critical ghost tourism, first and foremost, necessitates a commitment to social justice (Ateljevic et al., 2007, 3). This requires maintaining "an ethics of respect of the suffering a place has experienced," as Derek Dalton (2014, i) has put it. It is furthermore "imperative to learn something tangible about the history and legacy of that suffering." For tourists, this means recognizing the history of the land one occupies when traveling. Healing justice activists recommend "mapping conditions," which is useful here (Woodland, 2023d, 231). One should ask of the sites they are on and people they are visiting:

> What are the conditions of the land (environment)?
> What are the conditions of bodies (physical well-being)?
> What are the conditions of economies?
> What are the conditions of spirits (emotional, spiritual, psychic well-being)?
> What are the political and social conditions?

In our cotaught class, *Dark Dublin*, we have explored such questions when striving for for a critical ghost tour(ist) pedagogy. In the spring semester, we travel with students to Dublin, Ireland, for a one-week visit. We take students to the GPO museum where the Irish Volunteers and Citizen Army proclaimed the Irish Republic against British rule. This culminated into the 1916 Easter Rising and the eventual Irish War of Independence (1919–1921). We tour

Kilmainham Goal, where 14 leaders of the Easter Rising were executed in the prison's west wing. In the Dublin City Docklands, we visit the Famine Memorial, which commemorates the more than one million who died from starvation during the Great Famine (1845–1849). We furthermore interrogate the commercialization of violence, such as a tavern being named Darkey Kelly's Bar & Restaurant. The name refers to Dorcas Kelly who, upon being accused of witchcraft by Dublin Sheriff Simon Luttrell, was burned at the stake. Employee's T-shirts depict a hand engulfed in flames.

As a class, we, too, inquire on our collective fascination with the macabre—and how that may feed into the exploitation and spectatorship we seek to reject. As Patricia Morris and Tami Arford (2019, 422) assert,

> [W]e should engage in the difficult task of asking ourselves about our own complicity within a carceral landscape that continues to support systems of punishment which inflict suffering upon millions.

True crime, which certainly is an appendage to the dark tourism industry, for instance, often keeps consumers passive, voyeuristic, and in support of dominant modes of state violence (Felix & Garcia, 2023). Indeed, there is, as Saidiya Hartman (2022, 25) reminds us, a thin line between witness and spectator.

Critical ghost tourism also must consider the power imbalance between the tourist and the "toured." The tourist industry caters to those who are normatively able-bodied and who have the time and the financial means to travel (Bowman & Pezzullo, 2010, 197). In her writing on Caribbean tourism, Jamaica Kincaid (1998, 4–16) refers to the tourist, and of the white tourist in particular, as "an ugly human being": they are "a person visiting heaps of death and ruin and feeling alive and inspired at the sight of it." Kincaid speaks to the tourist's mindset: "[Y]ou needn't let that slightly funny feeling you have from time to time about exploitation, oppression, domination develop into full-fledged unease, discomfort; you could ruin your holiday."

We observed this dynamic in our travels. In Salem and New Orleans, we regularly walked past people begging on the streets or who appeared without shelter. A Salem guide remarked jokingly, "If someone is on the street yelling, which you have maybe seen, they may have a substance abuse or mental health problem, or they also may be possessed." We wrote during our walking tour in New Orleans, "Yesterday, our tour guide talked about bodies everywhere, in the sense of mass death and ghosts, yet we are actually walking around—and almost over—bodies of those unhoused." To be sure, the ghost tourism industry is an economic venture, funding cities divided on the basis of socioeconomic status. And the tourist is a consumer on this divide, which is navigated and negotiated by tour guides and by tourists themselves.

Finally, it is important to consider where the tourist returns: home, a place rife with violence, as Indigenous people, Black people, and immigrants have so often been displaced from home. As Kidada Williams (2023, 173) states for those terrorized by white vigilante night raids:

> Home would not let these families stay. It was now the shell of their burnt-out cabins or bullet-ridden houses. It was the place where the families had to sleep outside to avoid being violated again. Home was the place where their attackers were still roaming freely, either prowling around under the cloak of night or strutting proudly, boasting of their power, in the light of day.

In 1916, during Jim Crow segregation, Black families left the South en masse in what would be known as the Great Migration (Wilkerson, 2022, 279). They did so in search of a new home. As Emmett J. Scott (1920, 44) reported, "They were willing to make almost any sacrifice to obtain a railroad ticket, and they left with the intention of staying." On February 4, 1999, Liberian immigrant Amadou Diallo, who was making his home in the United States, was shot and killed by four New York police officers in front of a Bronx apartment. They fired 41 times (Armah, 2022).

To be sure, "home" is also rife with structural violence. As we write this in Pennsylvania, we are on land stolen from the Lenape People, and author Kevin has ancestral heritage to German settlement in Minnesota, a state where the hanging of 38 Dakota men occurred after the Dakota Wars (Berg, 2013). Pennsylvania has done much to erase the history of atrocity we have reviewed in this book. Indeed, the Lenape Nation of Pennsylvania continues to fight for state recognition, promoting petition- and letter-writing efforts to local and state legislators. Lawmakers, too, have initiated anti–critical race theory (CRT) legislation, with Republican Representatives introducing on June 7, 2021, the Teaching Racial and Universal Equality (TRUE) Act (HB 1532). The law would prohibit "critical race theory" (CRT) in school districts, state and local government entities, and public postsecondary institutions. Such legislation needs to be consistently challenged (see Revier, *forthcoming*).[11] Indeed, given that CRT focuses on interlocking forms of oppression (see Goldberg, 2023; Delgado & Stefancic, 2023), perhaps critical ghost tourism may be more aptly referred to as *critical race ghost tourism*.

Ultimately, a hauntologically informed examination of space (Fiddler, 2019, 474) demands a hauntologically informed sociological imagination (Mills, 1959; Gordon, 2008). This necessitates of the tourist, already in a troubled positionality, that they map conditions, both at home and abroad, and that they recognize America's horror stories of colonialism and slavery and their own varying complicity in upholding these legacies. This is especially so for whites who have largely maintained indifference, silence, and historical amnesia to the suffering of those racialized as not white. As Williams (2023, 220) reminds, "Short memories are the privilege of oppressors and their enablers."

In adopting these strategies, we can advocate for reparative justice, which requires *recognizing* structural harms and *committing* to financial and symbolic reparation "so that we may make something shared and new together from all the plundered parts" (Manjapra, 2022, 194). To engage in the past that continues to haunt us, we need to, as Erica Woodland (2023b, 209) contends,

> [S]imultaneously hold the past and present horrors of the state, along with the sacrifices of those who came before us, with the radical possibilities that we can build the liberated systems of care we need.

A *spirited Black feminist hauntology* (Saleh-Hanna, 2015)—which recognizes and reckons with the America's past-present-future horrors through the metaphor of the ghost(ed)—is key to progress toward justice, inside and outside of the ghost tourism industry.

Notes

1 Williams (2023) reviews the congressional report for the 1871–1872 Klan hearings and interviews conducted by the Works Progress Administration (WPA) in the 1930s (Williams, 2023, xxii).
2 While historical records afford a larger scope of America's horror stories, we must nevertheless be cognizant of, as Saidiya Hartman (2022, 12) reminds, "the impossibility of fully recovering the experience of the enslaved and the emancipated, and the risk of reinforcing the authority of these documents"
3 It is more of a symbolic birthday, as there is evidence that Africans may have joined Spanish expeditions to the present-day United States during the 16th century. Africans also prevented Spanish slaveholders from establishing plantations in 1526 in what is now South Carolina. A muster roll also shows that there were already 32 African slaves in Virginia in March 1619 (Holt, 2011, 3). As Kendi (2022, xiv) reminds us, "No one knows how or when they arrived. No one knows the *precise* birthdate of African America." August 20, 1619 therefore offers a historically documented, symbolically meaningful touch point.
4 We have to shout out speculative fiction writer and YouTuber Krishana Davis (2022) for recommending these readings in her video, "10 of the BEST Thriller, Horror Books by Black Authors | My Recommendations for Good Reading."
5 The 1968 WITCH Manifesto (Women's International Terrorist Conspiracy from Hell) proclaims, "There is no 'joining' WITCH. If you are a woman and dare to look within yourself, you are a Witch."
6 They are, however, critical of the medical industrial complex (MIC) and of its connections with the prison industrial complex (PIC) as they both uphold racist population control (Page, 2023).
7 Healing justice operates through three basic principles: collective trauma is transformed collectively; there is no single model of care; and healing strategies are rooted in place and ancestral technologies (Middleton & Page, 2023, 126). As such, they promote variegated environmental, reproductive, and transformative justices as well as liberatory harm reduction (Page & Woodland, 2023a).
8 This portion of the statement is worth citing in whole given the breadth of focus by the Combahee River Collective,

The most general statement of our politics at the present time would be that we are actively committed to struggling against racial, sexual, heterosexual, and class oppression, and see as our particular task the development of integrated analysis and practice based upon the fact that the major systems of oppression are interlocking. The synthesis of these oppressions creates the conditions of our lives. As Black women we see Black feminism as the logical political movement to combat the manifold and simultaneous oppressions that all women of color face.

9 It should be noted that the historical foundation of white supremacy is based on ideals surrounding Anglo-Saxon history and heritage (Strings, 2019, 129).
10 He does so when referring to how the U.S. Constitution never mentions slavery.
11 Lawmakers evoke the words of Dr. Martin Luther King Jr.'s "I Have a Dream Speech," where he states, "I have a dream that my four little children will one day live in a nation where they will not be judged by the color of their skin but by the content of their character." CRT, they conclude, is "exactly the opposite of Dr. King's dream."

References

Adamkiewicz, Ewa A. 2016. White Nostalgia: The Absence of Slavery and the Commodification of White Plantation Nostalgia. Aspeers: Emerging Voices in American Studies, 9.1:13–31.

American Indian Movement. 1972, October. "Trail of Broken Treaties 20-Point Position Paper." Utah State University. https://www.usu.edu.

Amsden, David. 2015, February 26. "Building the First Slavery Museum in America." New York Times Magazine. https://www.nytimes.com.

Armah, Esther. 2022. "The Black Immigrant." In Four Hundred Souls: A Community History of African America, 1619–2019, edited by Ibram X. Kendi and Keisha N. Blain, 370–373. New York: One World.

Ateljevic, Irena, Nigel Morgan, and Annette Pritchard. 2007. "Editors' Introduction: Promoting an Academy of Hope in Tourism Enquiry." In The Critical Turn in Tourism Studies: Innovative Research Methodologies, edited by Irena Ateljevic, Annette Pritchard, and Nigel Morgan, 1–8. Oxford: Elsevier.

Barber II, William J. 2022. "David George." In Four Hundred Souls: A Community History of African America, 1619–2019, edited by Ibram X. Kendi and Keisha N. Blain, 135–138. New York: One World.

Bell, Derrick A. 1987. And We Are Not Saved: The Elusive Quest for Racial Justice. New York: Basic Books.

Berg, Scott W. 2013. Nooses: Lincoln, Little Crow, and the Beginning of the Frontier's End. New York: Vintage.

Blackmon, Douglas A. 2009. Slavery by Another Name: The Re-Enslavement of Black Americans from the Civil War to World War II. Sioux City: Anchor.

Bowman, Michael S., and Phaedra C. Pezzullo. 2009. "What's So 'Dark' about 'Dark Tourism'?: Death, Tours, and Performance." Tourist Studies, 9(3): 187–202.

Boyd, Herb. 2022. "The Revolt in New York." In Four Hundred Souls: A Community History of African America, 1619–2019, edited by Ibram X. Kendi and Keisha N. Blain, 82–84. New York: One World.

Breslaw, Elaine G. 1995. Tituba, Reluctant Witch of Salem: Devilish Indians and Puritan Fantasies. New York: New York University Press.

Broaddus, Maurice. 2023. "The Norwood Trouble." In Out There Screaming: An Anthology of New Black Horror, edited by Jordan Peele, 225–240. New York: Penguin.

Brooks, Robin. 2018. "RIP Shirts or Shirts of the Movement: Reading the Death Paraphernalia of Black Lives." Biography, 41(4): 807–830.

Brown, Jericho. 2022. "Upon Arrival." Four Hundred Souls: A Community History of African America, 1619–2019, edited by Ibram X. Kendi and Keisha N. Blain, 34–35. New York: One World.

Cariou, Warren. 2006. "Haunted Prairie: Aboriginal 'Ghosts' and the Specters of Settlement." University of Toronto Quarterly, 75(2): 727–734.

Chavez, Roby. 2022. "New Orleans Was Once the Center of U.S. Slave Trade. This Artist Wants to Make Sure We Don't Forget." PBS. https://www.pbs.org.

Chollet, Mona. 2022. In Defense of Witches: The Legacy of the Witch Hunts and Why Women Are Still on Trial. New York: St. Martin's Press.

Clark, P. Djèlí. 2020. Ring Shout. New York: Tordotcom.

Cohen, Stanley. 2013. States of Denial: Knowing about Atrocities and Suffering. New Jersey: John Wiley & Sons.

Combahee River Collective. 1977, April. "Combahee River Collective Statement." https://combaheerivercollective.weebly.com/the-combahee-river-collective-statement.html.

Condé, Maryse. 1992. I, Tituba, Black Witch of Salem. Charlottesville: University of Virginia Press.

Crenshaw, Kimberlé. 2024. # SayHerName: Black Women's Stories of Police Violence and Public Silence. Chicago: Haymarket Books.

Dalton, Derek. 2014. Dark Tourism and Crime. New York and London: Routledge.

Davis, Angela Y. 2003. Are Prisons Obsolete? New York: Seven Stories Press.

Davis, Krishana. 2022. "10 of the BEST Thriller, Horror Books by Black Authors | My Recommendations for Good Reading." YouTube. https://www.youtube.com.

Delgado, Richard, and Jean Stefancic. 2023. Critical Race Theory: An Introduction (4th ed.). New York: New York University Press.

Deyab, Mohammad Shaaban Ahmad. 2016. "Cultural Hauntings in Toni Morrison's Beloved (1987)." English Language, Literature & Culture, 1(3): 13–20.

Diouf, Sylviane A. 2022. "Maroons and Marronage. Prince." In Four Hundred Souls: A Community History of African America, 1619–2019, edited by Ibram X. Kendi and Keisha N. Blain, 89–92. New York: One World.

Douglas, Deborah. 2022. "Hurricane Katrina." In Four Hundred Souls: A Community History of African America, 1619–2019, edited by Ibram X. Kendi and Keisha N. Blain, 374–377. New York: One World.

Du Bois, W.E.B. 1935. Black Reconstruction in America: An Essay Toward a History of the Part Which Black Folk Played in the Attempt to Reconstruct Democracy in America, 1860–1880. A. Saifer.

Due, Tananarive. 2023. "The Rider." In Out There Screaming: An Anthology of New Black Horror, edited by Jordan Peele, 97–115. New York: Penguin.

Eastern State Penitentiary. "The Big Graph." https://www.easternstate.org/explore/exhibits/big-graph.

Faust, Drew Gilpin. 2008. This Republic of Suffering: Death and the American Civil War. New York: Vintage.

Federici, Silvia. 2018. Witches, Witch-Hunting, and Women. Binghamton: PM Press.

Felix, Shanna N., and Merideth Garcia. 2023. "The Hauntings of Kitty Genovese: The Bystander Effect and Queer Invisibility." In The (Mis) Representation of Queer Lives in True Crime, edited by Abbie E. Goldberg, Danielle C. Slakoff, and Carrie L. Buist, 141–159. New York and London: Routledge.

Fiddler, Michael. 2019. "Ghosts of Other Stories: A Synthesis of Hauntology, Crime and Space." Crime, Media, Culture, 15(3): 463–477.

Fiddler Michael, Linnemann, Travis, and Theo Kindynis. 2024. "Ghost Criminology: A Framework for the Discipline's Spectral Turn." British Journal of Criminology, 64(1): 1–16.

Foley, Laura. 2021. "The Haunted History of New Orleans: An Exploration of the Intersectionality between Dark Tourism, Black History, and Public History." ProQuest Dissertation Publishing. Glassboro: Rowan University.

Fry, Gladys-Marie. 2001. Night Riders in Black Folk History. Chapel Hill, NC: University of North Carolina Press Books.

Garza, Alicia. 2022. "Black Lives Matter." In Four Hundred Souls: A Community History of African America, 1619–2019, edited by Ibram X. Kendi and Keisha N. Blain, 382–386. New York: One World.

Gibson, Marion. 2024. Witchcraft: A History in Thirteen Trials. New York: Simon & Schuster.

Gilmore, Ruth Wilson. 2007. Golden Gulag: Prisons, Surplus, Crisis, and Opposition in Globalizing California. Berkeley: University of California Press.

Goldberg, David Theo. 2023. The War on Critical Race Theory: Or, the Remaking of Racism. Hoboken: John Wiley & Sons.

Gordon, Avery F. 2008. Ghostly Matters: Haunting and the Sociological Imagination. Minneapolis: University of Minnesota Press.

Gordon, Avery F. 2011. "Some Thoughts on Haunting and Futurity." Borderlands, 10(2): 1–21.

Gumbs, Alexis Pauline. 2022. "Phillis Wheatley." In Four Hundred Souls: A Community History of African America, 1619–2019, edited by Ibram X. Kendi and Keisha N. Blain, 130–134. New York: One World.

Gumbs, Alexis Pauline. 2023. "Learning to listen." In Healing Justice Lineages: Dreaming at the Crossroads of Liberation, Collective Care, and Safety, edited by Cara Page and Erica Woodland, 15–20. Berkeley: North Atlantic Books.

Hannah-Jones, Nikole. 2022. "Arrival." In Four Hundred Souls: A Community History of African America, 1619–2019, edited by Ibram X. Kendi and Keisha N. Blain, 3–7. New York: One World.

Harriet's Apothecary. 2024. "Who We Are." https://www.harrietsapothecary.com.

Hartman, Saidiya. 2022. Scenes of Subjection: Terror, Slavery, and Self-Making in Nineteenth-Century America. New York: W.W. Norton & Company.

Hayes, Kelly, and Mariame Kaba. 2023. Let This Radicalize You: Organizing and the Revolution of Reciprocal Care. Chicago: Haymarket Books.

Holloway, Karla F.C. 2003. Passed On: African American Mourning Stories, A Memorial. Durham: Duke University Press.

Holt, Thomas C. 2011. Children of Fire: A History of African Americans. New York: Hill and Wang.

Horwitz, Tony. 1998. Confederates in the Attic: Dispatches from the Unfinished Civil War. New York: Vintage.

Jackson, Kellie Carter. 2022. "The Virginia Law That Forbade Bearing Arms; or the Virginia Law That Forbade Armed Self-Defense." In Four Hundred Souls: A Community History of African America, 1619–2019, edited by Ibram X. Kendi and Keisha N. Blain, 55–56. New York: One World.

James, Elizabeth. 2014. African American Informal Focus Group, New Orleans.

Kaba, Mariame. 2019. "Black women Punished for Self-Defense Must Be Freed from Their Cages." The Guardian. https://www.theguardian.com.

Kendi, Ibram X. 2022. "A Community of Souls: An Introduction." In Four Hundred Souls: A Community History of African America, 1619–2019, edited by Ibram X. Kendi and Keisha N. Blain, xiii–xvii. New York: One World.

Kendi, Ibram X., and Keisha N. Blain. 2022. Four Hundred Souls: A Community History of African America, 1619–2019. New York: One World.

Kincaid, Jamaica. 1988. A Small Place. New York: Farrar, Straus and Giroux.

Kitwana, Bakari. 2022. "The Hip-Hop Generation." In Four Hundred Souls: A Community History of African America, 1619–2019, edited by Ibram X. Kendi and Keisha N. Blain, 355–358. New York: One World.

Lenape Nation of Pennsylvania. 2018. "Current Stand for State Recognition." https://www.lenape- nation.org.

Long, Carolyn Morrow. 2007. A New Orleans Voudou Priestess: The Legend and Reality of Marie Laveau. Gainesville: University Press of Florida.

"Louisiana: A Guide to the State." 1941. Louisiana Library Commission. New York: Hastings House.

Love, David A. 2022. In Four Hundred Souls: A Community History of African America, 1619–2019, edited by Ibram X. Kendi and Keisha N. Blain, 47–50. New York: One World.

Lowenthal, David. 2015. The Past Is a Foreign Country, Revisited (2nd ed.). Cambridge: Cambridge University Press.

Lowery, Wesley. 2022. "The Stono Rebellion." In Four Hundred Souls: A Community History of African America, 1619–2019, edited by Ibram X. Kendi and Keisha N. Blain, 111–114. New York: One World.

Lucie, Armitt. 2000. Contemporary Women's Fiction and the Fantastic. New York: St Martin's Press.

Manjapra, Kris. 2022. Black Ghost of Empire: The Long Death of Slavery and the Failure of Emancipation. New York: Simon & Schuster.

Martinez, Aja Y. 2020. Counterstory: The Rhetoric and Writing of Critical Race Theory. Champaign: National Council of Teachers of English.

McGhee, Heather C. 2022. "Bacon's Rebellion." In Four Hundred Souls: A Community History of African America, 1619–2019, edited by Ibram X. Kendi and Keisha N. Blain, 51–54. New York: One World.

McWilliams, Kelly. 2023. Your Plantation Prom Is Not OK. New York: Hachette Book Group.

Middleton, Tamika, and Cara Page. 2023. "Conjuring the Roots of Healing Justice in the Southeast." In Healing Justice Lineages: Dreaming at the Crossroads of Liberation, Collective Care, and Safety, edited by Cara Page and Erica Woodland, 119–132. Berkeley: North Atlantic Books.

Miles, Tiya. 2015. Tales from the Haunted South: Dark Tourism and Memories of Slavery from the Civil War Era. Chapel Hill: University of North Carolina Press.

Miles, Tiya. 2022, November 14. "In Anxious Times, Black History Can Be a Blueprint for Survival." New York Times.

Mills, C. Wright. 1959. The Sociological Imagination. Oxford: Oxford University Press.

Morgan, Jennifer L. 2022. "Elizabeth Keye." In Four Hundred Souls: A Community History of African America, 1619–2019, edited by Ibram X. Kendi and Keisha N. Blain, 39–42. New York: One World.

Morris, Patricia, and Tammi Arford. 2019. "'Sweat a Little Water, Sweat a Little Blood': A Spectacle of Convict Labor at an American Amusement Park." Crime, Media, Culture, 15(3): 423–446.

Morrison, Toni. 1987. Beloved. New York: Vintage Books.

New Orleans Slave Trade. https://www.neworleansslavetrade.org.

O'Donovan, Maria. 2019. "Nostalgia and Heritage in the Carousel City: Deindustrialization, Critical Memory, and the Future." Journal of Community Archaeology & Heritage, 6(4): 272–282.

Oluo, Ijeoma. 2022. "Whipped for Lying with a Black Woman." In Four Hundred Souls: A Community History of African America, 1619–2019, edited by Ibram X. Kendi and Keisha N. Blain, 11–14. New York: One World.

Owens, Deirdre Cooper. 2022. "The Fugitive Slave Act." In Four Hundred Souls: A Community History of African America, 1619–2019, edited by Ibram X. Kendi and Keisha N. Blain, 162–165. New York: One World.

Page, Cara. 2023. "Roots of the Medical Industrial Complex." In Healing Justice Lineages: Dreaming at the Crossroads of Liberation, Collective Care, and Safety, edited by Cara Page and Erica Woodland, 56–65. Berkeley: North Atlantic Books.

Page, Cara, and Erica Woodland. 2023a. Healing Justice Lineages: Dreaming at the Crossroads of Liberation, Collective Care, and Safety. Berkeley: North Atlantic Books.

Page, Cara, and Erica Woodland. 2023b. "Introduction." In Healing Justice Lineages: Dreaming at the Crossroads of Liberation, Collective Care, and Safety, edited by Cara Page and Erica Woodland, 1–9. Berkeley: North Atlantic Books.

Page, Cara, and Erica Woodland, 2023c. "Conclusion." In Healing Justice Lineages: Dreaming at the Crossroads of Liberation, Collective Care, and Safety, edited by Cara Page and Erica Woodland, 265–272. Berkeley: North Atlantic Books.

Parker, Nakia D. 2022. "Unfree Labor." In Four Hundred Souls: A Community History of African America, 1619–2019, edited by Ibram X. Kendi and Keisha N. Blain, 30–33. New York: One World.

Peele, Jordon. 2023. Out There Screaming: An Anthology of New Black Horror. New York: Penguin.

Perry, Imani. 2022. "Black Arts." In Four Hundred Souls: A Community History of African America, 1619–2019, edited by Ibram X. Kendi and Keisha N. Blain, 321–324. New York: One World.

Powell, John A. 2022. "Dred Scott." In Four Hundred Souls: A Community History of African America, 1619–2019, edited by Ibram X. Kendi and Keisha N. Blain, 214–217. New York: One World.

Ralph, Laurence. 2022. "The Code Noir." In Four Hundred Souls: A Community History of African America, 1619–2019, edited by Ibram X. Kendi and Keisha N. Blain, 57–61. New York: One World.

Revier, Kevin. "Flames of (In)Justice: A (White) Book-Burning Parable on Anti-Critical Race Theory Legislation." Writers: Craft and Context, 5(1). Forthcoming.

Roberts, Dorothy E. 2022. "Race and the Enlightenment." In Four Hundred Souls: A Community History of African America, 1619–2019, edited by Ibram X. Kendi and Keisha N. Blain, 119–122. New York: One World.

Saleh-Hanna, Viviane. 2015. "Black Feminist Hauntology. Rememory the Ghosts of Abolition?" Champ Pénal/Penal Field, 12.

Scott, Emmett J. 1920. Negro Migration during the War. New York: Oxford University Press.

Seidule, Ty. 2021. Robert E. Lee and Me: A Southerner's Reckoning with the Myth of the Lost Cause. New York: St. Martin's Press.

Singh, Amardeep. 2021. "The Story of Margaret Garner: Inspiration for 'Beloved.'" Lehigh University.

Smith, Barbara. 2022a. "The Combahee River Collective." In Four Hundred Souls: A Community History of African America, 1619–2019, edited by Ibram X. Kendi and Keisha N. Blain, 340–343. New York: One World.

Smith, Clint. 2022b. "The Louisiana Rebellion." In Four Hundred Souls: A Community History of African America, 1619–2019, edited by Ibram X. Kendi and Keisha N. Blain, 173–176. New York: One World.

Soloman, Rivers. 2017. Unkindness of Ghosts. New York: Akashic Books, Ltd.

Stevenson, Bryan. 2017. "A Presumption of Guilt: The Legacy of America's History of Racial Injustice." In Policing the Black Man, edited by Angela J. Davis, 3–30. New York: Pantheon Books.

Stevenson, Bryan. 2021. "Punishment." In The 1619 Project: A New Origin Story, edited by Nikole Hannah-Jones, 275–283. New York: One World.

Stevenson, Brenda E. 2022. "Black Women's Labor." In Four Hundred Souls: A Community History of African America, 1619–2019, edited by Ibram X. Kendi and Keisha N. Blain, 18–21. New York: One World.

Strings, Sabrina. 2019. Fearing the Black Body: The Racial Origins of Fat Phobia. New York: New York University Press.

Taylor, Terence. 2023. "Your Happy Place." In Out There Screaming: An Anthology of New Black Horror, edited by Jordan Peele, 306–322. New York: Penguin.

Thompson-Spires, Nafissa. 2022. "Lucy Terry Prince." In Four Hundred Souls: A Community History of African America, 1619–2019, edited by Ibram X. Kendi and Keisha N. Blain, 115–118. New York: One World.

Tisby, Jemar. 2022. "The Virginia Law on Baptism." In Four Hundred Souls: A Community History of African America, 1619–2019, edited by Ibram X. Kendi and Keisha N. Blain, 43–46. New York: One World.

Turner, Sasha. 2022. "The Slave Market." In Four Hundred Souls: A Community History of African America, 1619–2019, edited by Ibram X. Kendi and Keisha N. Blain, 85–88. New York: One World.

Utah, Adaku, and Cara Page. 2023. "A Constellation of Healing Justice and Liberation: Sites of Practice in New York City." In Healing Justice Lineages: Dreaming at the Crossroads of Liberation, Collective Care, and Safety, edited by Cara Page and Erica Woodland, 176–183. Berkeley: North Atlantic Books.

Waldstreicher, David. 2010. Slavery's Constitution: From Revolution to Ratification. New York: Hill and Wang.

Walker, Corey D.B. 2022. "The Spirituals." In Four Hundred Souls: A Community History of African America, 1619–2019, edited by Ibram X. Kendi and Keisha N. Blain, 93–95. New York: One World.

Washington, Harriet A. 2008. Medical Apartheid: The Dark History of Medical Experimentation on Black Americans from Colonial Times to the Present. New York: Doubleday Books.

Wells, Ida B. 1982. Southern Horrors: Lynch Law in All Its Phases.

Wilkerson, Isabel. 2022. "The Great Migration." In Four Hundred Souls: A Community History of African America, 1619–2019, edited by Ibram X. Kendi and Keisha N. Blain, 93–95, 278–282. New York: One World.

Williams, Kidada E. 2023. I Saw Death Coming: A History of Terror and Survival in the War against Reconstruction. New York: Bloomsbury Publishing.

Woodland, Erica. 2023a. "Uninterrupted Legacy of Resistance." In Healing Justice Lineages: Dreaming at the Crossroads of Liberation, Collective Care, and Safety, edited by Cara Page and Erica Woodland, 32–42. Berkeley: North Atlantic Books.

Woodland, Erica. 2023b. "We Are Our History: Transforming the Legacy of Criminalization." In Healing Justice Lineages: Dreaming at the Crossroads of Liberation, Collective Care, and Safety, edited by Cara Page and Erica woodland, 66–85. Berkeley: North Atlantic Books.

Woodland, Erica. 2023d. "Storytelling Practice: Our Land, Work, Bodies, and Spirit." In Healing Justice Lineages: Dreaming at the Crossroads of Liberation, Collective Care, and Safety, edited by Cara Page and Erica Woodland, 133–14. Berkeley: North Atlantic Books.

Wright, Kai. 2022. "The Virginia Slave Codes." In Four Hundred Souls: A Community History of African America, 1619–2019, edited by Ibram X. Kendi and Keisha N. Blain, 77–78. New York: One World.

BIBLIOGRAPHIC INDEX

SUBJECT INDEX